NUT FRUITS FOR THE HIMALAYAS

Concept's Discovering Himalayas Series No. 2

Nut Fruits for The Himalayas

M.L. DEWAN
M.C. NAUTIYAL
V.K. SAH

Foreword by
K.L. CHADHA
Deputy Director General (Hort.)
Indian Council of Agricultural Research,
Krishi Bhawan, New Delhi, India

CONCEPT PUBLISHING COMPANY, NEW DELHI 110059

Sponsored By
Trees for Life, India

ISBN 81-7022-399-7

First Published 1992

Published and Printed by
Ashok Kumar Mittal
Concept Publishing Company
A/15-16, Commercial Block, Mohan Garden
NEW DELHI-110059 (India)

Lasertypeset by
Microtech Advance Printing Systems (Pvt.) Ltd.
H-13, Bali Nagar
NEW DELHI 110015

Dedicated to

THE SMALL FARMERS

and

FARM WOMEN

of

THE HIMALAYAS

Whose Involvement and Perservence

Can Make Success of

The Himalayan Nut Fruit Programme

FOREWORD

NUT CROPS are very important crops of the temperate region and include walnut, chestnut, almond, pecannut, hazelnut, apricot, pistachio nut and chilgoza. These are rich sources of vitamins, fats, proteins and minerals and are highly prized for their growth promoting activities in human beings. Tree nuts also help in conserving the soil due to their deep root system.

Although horticulture has made rapid strides in the North West Himalayan region, nut cultivation has not received the attention it deserved. Cultivation of crops like walnut and chilgoza can boost our export earnings while cultivation of crops like pistachio and almond can help us in conserving our foreign exchange. Besides this, such low volume high value crops can be more profitable particularly in far flung areas with inadequate transport means and can help in ecological stability of the region.

Efforts are underway to increase nut production in the country and the present book is an effort in the right direction. The book describes the scientific cultivation of all important nut fruits namely walnut, almond, chestnut, hazelnut, pecan nut, pistachio nut, chilgoza nut, apricot, macadamia nut and other minor nuts. Various aspects of each crop covering area and production, varieties, propagation methods, nutritional requirements and control of pests and diseases are well presented in the book.

Various production constraints have been identified in each nut crop and future lines of work suggested. The book is the outcome of deep interest which Dr. M.L. Dewan the senior author has in the conservation of Himalayas.

The authors have done a commendable job in compiling the available literature on nut crops in the country and abroad in the present book entitied *Nut Fruits for the Himalayas*. The book is well written and well illustrated. I congratulate the authors for their efforts in writing this book and "Trees for life, India" organization for sponsoring this noble cause. I am sure, the book will prove very useful to orchardists, researchers and extension agencies particularly in North West Hill Region of the country.

K.L. CHADHA
Deputy Director General (Hort)
August 23, 1991 ICAR, Krishi Bhawan, New Delhi

ACKNOWLEDGEMENTS

NUTS FRUITS for the Himalayas, owes much to the help, and prepared under "Concept's Discovering Himalayas Series" sponsored by Trees for Life, cooperation and inspiration received from a large number of people, voluntary agencies and the industry. Our grateful thanks to Shri Darbari Seth, Chairman, Tata Tea Limited, who encouraged us to conceive and execute the project of greening the Himalayas, to conserve its ecology and stop further degradation. A long-range planning of planting walnuts, almonds, chestnuts, etc., in the area is the foresight of the authors.

Grateful thanks are also due to Shri Balbir Mathur, a Non-Resident Indian in the USA who is the founder of Trees for Life, a voluntary charitable organization planning to green the countryside through a revolutionary fruit-tree planting project and alleviate hunger, and Mr. Sharad Chaturvedi, Executive Director for India Chapter of TFL, and a number of voluntary organizations in the Central Himalayas which have taken upon themselves this task of promoting nuts as an income-generating activity of the Himalayan community.

Our grateful thanks also go to Dr. K L Chadha, Deputy Director-General, ICAR, New Delhi, and Dr. R M Pandey of the National Horticulture Board for their encouragement in formulating such a programme.

Thankful mention is being made of a friend of

ours, Shri A D Verma, formerly of the Planning Commission, who helped put the material for this publication together in the early stages of preparation of the text for the book. Our thanks are also due to Dr. P L Gautam, Joint Director, Hill Campus, G.B. Pant Agriculture University, Ranichauri, and Dr. J. Kumar, Sr. Scientist Pathology who were helpful during the preparation of the manuscript.

Shri Sona Ullah Banihali of Banihal, Jammu and Kashmir, a politician turned environmentalist and Vriksha Mitra awardee, greatly helped in acquiring/collecting high quality seeds of walnuts and other nuts from Kashmir and training our farmers in the U.P. Himalayas. Our sincere thanks go to him as also to his co-workers Shri Avinash Bhakri and Shri Asadullah.

We will be failing in our duty if we do not duly thank the various voluntary agencies and men behind them who enthused and helped us in this programme. They are Shri Yogesh Chandra Bahuguna of Parvateeya Jan Kalyan Samiti, Ranichauri; Shri Pratap Shekhar of Uttarakhand Jan Jagriti Samiti, Khadi Jajal; Sarvashri Kanai Lal and Kalyan Paul of Chirag, Sitla; Shri Sadan Mishra of Berinag Gram Swarajya Mandal, Kandey, Pithoragarh; Smt Savitri Bahen of Sabli Mahila Mandal; and Shri Ganga Prasad Bahuguna, a social worker at Ranichauri.

Lastly, we are once again thankful to all these people who have put their heart in the effort for the conservation and development of the Himalayas and the Himalayan community, and without whose active cooperation the production of this manual, could not have been possible.

AUTHORS

CONTENTS

LIST OF PLATES

LIST OF TABLES

1

INTRODUCTION

Hunger knows no friend but its feeder.
—Aristophane

MAN'S demand for food from the natural resources that sustain his very existence has increased enormously in recent years. Seeing the present growth rate it may be expected that the world's population by the year A.D. 2000 will cross the six billion mark putting double or triple pressure on the earth for meeting the enormously increased demand for food and agricultural products. But the land resources at global level will be tragically inadequate to meet the desired production level to feed the massive future populations.

In the developing countries an unprecedented increase in the population has placed a tremendous pressure on the natural resources resulting in their rapid degradation or depletion, particularly soil, water and vegetation. Increasing demand for food and agricultural products has become a prime need to solve the food problem of the human and livestock population.

In the Himalayan context, conservation of the natural resources particularly soil and water is of great importance, and has to be undertaken with due care keeping in view the fragile eco-system. The land resources are to be used for boosting of the food production of high nutritive value, so that the problems like starvation and malnutrition are solved.

In this, tree-nuts cultivation in the Himalayan mountain region will greatly help in utilizing the land resources effectively on the one hand and will solve the socio-economic problems of the zone on the other. Among the major tree nuts which have been cultivated through the centuries as a rich source of food are, walnut, chestnut, almond, pecan, hazelnut, chilgoza, etc., in the region. These have been described scientifically, based on the experiences of many progressive orchardists, researchers, extension workers and nursery experts. In the modern times, an average of 281,000 metric tonnes tree nuts are produced worldwide, out of which 200,000 tonnes approximately of edible kernels are used either as a food source or for industrial purposes, being a rich source of fats, protein, vitamins and minerals. Besides their nutritive value tree nuts help in conserving the soil due to their deep and well-expanded root system and help in regulating the nutrient and hydrological cycle.

It is of major significance that the hill communities now recognize the necessity of altering the patterns of landuse in the north-west Himalayas to maintain an equilibrium between socio-economic and eco-geographic problems. Ecological problems of the mountanous tracts are quite complex due to continuous denudation and depletion of the vegetation wealth particularly on the fragile and steep slopy terrains. This has not only resulted in low productivity of the conventional cereal crops being grown in the area but has also shown devastating effects like soil erosion, drought and flood, making the subsistence a challenge for the hill communities. Improved agro-technology being input-intensive has not helped much, partly perhaps between lack of harmony between the farming communities and the developmental agencies. Keeping the entire picture in view the existing landuse needs to be altered and replaced by tree crops which will be more remunerative as compared to cereal crops. This will be helpful in maintaining the ecological balance as well as in improving the socio-economic status of the hill community. Participation of people, voluntary organizations and Government or semi-Government organizations together can bring about socio-economic and eco-geographical revolutions here in the true sense. Perishable fruits have also not proved a more paying proposition for the hill communities, the reason being lack of proper storage and marketing infrastructure. In the present context popularization of a nut production programme should prove a step in the right direction leading to improved income and ecological stability. The programme can be made more useful and popular, if the following considerations are taken care of :

— A very bold attempt needing well-planned and sustainable strategy and viable organizational, infrastructural and institutional support. Research and Development is pre-requisite in view of the long life span of nut-producing trees.

— The hill community particularly the economically weaker section, school children and dropouts, unemployed youth, women folk, voluntary organizations should be actively involved in the programme and strong research support provided continuously till the programme is made widely acceptable.

NGO's thrust such as from the Trees for Life and others is very good but as the scope of nut extension often lies in far remote areas where communication is still poor, promotion needs establishment of suitably organized channels of extension agencies. Establishment of demonstration/seed production orchards, need scientific inputs, trained manpower, adequate extension literature. Universities and other institutions should be involved in the programme.

— Since the north-west Himalayas offer tremendous diversity in agro-climatic conditions, plans for growing specific nut crops for a particular ecological situation need to be formulated. Acceptance of the crop in view of the local community need is to be promoted. Local strains of nut crops must be given priority and continuous efforts for their improvement should essentially be made. Plant introduction from one to the other area of a states or from State to State is a critical factor. Local strains/cultivars have higher establishment possibilities. Very careful quarantine control on seed import is also crucial. The planting targets, therefore, should not be the major plank of the programme, rather local success stories should be developed into future strategy.

— Development of a new or expanded nut industry technology to diversify landuse must be made feasible. Limitations can be biological, technological, economic or sociological. All aspects are important from one angle or the other. Technical and biological factors need research and development to give confidence to land users. Social barriers to landuse change are very real and, without community support, attempts to initiate such programmes will not give adequate results.

— As per existing practice in conventional nut

production, weak and undersized nursery stocks are being used for out-planting, resulting in poor survival. Therefore, in order to achieve higher success quality planting stock with the desired height should invariably be used.

— Setting up of high quality demonstration nurseries/production nurseries are absolutely essential. Technical inputs into this are of much importance as in the quality of seed. With requisite inputs and dedication, it is possible to produce sufficient quality nursery stock in a shorter period of 12 to 18 months instead of looking for a plantable seedling in 24 to 36 months. For example, selection of dwarf and early maturing walnut is a task of high scientific order. The moisture and space requirements of a majority of the nut species are exacting and the choice of soil and site in the small holdings of the Himalayan peasantry need careful examination. Small holdings can often be a deterrent to farmers' interest in growing nuts.

— Every nut species requires very specific climatic conditions which differ from place to place in accordance with the topographical and exposure to the sun, rain and wind (micro-climate) features of the area. Hence, prior to implementing a nut production programme with any specific species and their respective cultivars, such factors need to be critically examined so as to ensure the profitability of crop in the long run. Likewise, within the Himalayas, the outer Himalayas are climatically entirely different from the inner Himalayan hills of Uttar Pradesh and Himechal Pradesh, and therefore, the outer Himalayan tracts can be suitably selected for the edible nuts like Pecan nut, low chilling Almond, Macadamia, Cheora, etc., demanding a fairly warmer climate and higher rainfall. Inner Himalayan hills have a rather high potential for Walnut, Chestnut, Almonds and Chilgoza Pine nut cultivation.

To sum up, the value of nut trees in a farming situation depends upon the inherent potential of that environment, the kind of prevailing climate of the area, the genetic material available and the indirect and direct use or value of the trees and their products. However, the purpose and need for which nut trees are grown is of prime importance and it should be able to answer the problems like economic stability, feasibility, social acceptance and its long-term effect on the overall ecology of the zone.

In addition, there is need for much greater scientific appreciation of the environment to make meaningful decisions on many issues for large-scale nut production. These issues are: involvement of local people for adopting an alternative landuse for environmental and socio-economic improvement, motivation and creating awareness among the farming community that such a programme has a great potential for income generation. These are essential for such programme to achieve real success.

The present work deals with the crop management technology, promising cultivars, pollination, fruiting behaviour, etc., of important edible nuts for involving the rural hill peasantry in the programme entitled "Promotion and Production of Edible Nuts in the Himalayas", which has been sponsored by "Trees for Life", India. The work is intended to meet the need for a handy reference and working manual. It has been written simply, but contains adequate technical material to serve the purposes of hill orchardists, students, NGOs, grassroot extension workers and other interested institutions. It is hoped it will help in expanding the area under tree nuts in the Himalayan mountains for solving the multifaceted problem of food, timber, fuel and environment, and will go a long way in improving the socio-economic conditions of the Himalayan communities.

2

PRODUCTION AND PROMOTION OF EDIBLE NUTS IN THE HIMALAYAS

WOODMAN, SPARE THAT TREE !
TOUCH NOT A SINGLE BOUGH !
IN YOUTH IT SHELTERED ME,
AND I'LL PROTECT IT NOW.

—*George Pope Morris*

AGRICULTURE and horticultural introduction in the Himalayas along with conservation and tree-planting and water-harvesting are becoming more and more relevant in the ecological development of the Himalaya. Horticulture, particularly the temperate fruits, is highly suited to many parts of the Himalayas but, due to geographical constraints and poor infrastructure like lack of roads and communications, insufficient marketing and storage facilities, the perishable and semi-perishable fruits do not fetch a proper price for the farmer.

The production of nuts like, walnut, almond, chestnut, hazelnut, etc., which have a much longer lasting life, can therefore, open avenues of increased nutrition and income for the Himalayan people. There is a great need for such promotion in the Uttar Pradesh Himalayas – eight hill districts. Such type of promotion has been experienced in Jammu and Kashmir (J&K) and started in Himachal Pradesh. A beginning has been made under a Nuts

Project, where some seeds have been obtained from J&K nurseries and raising of quality stock has been initiated in order to achieve high quality nuts. For a proper evaluation of nut strains plant materials need to be imported from countries like Italy, Turkey, Iran, Afghanistan, etc. The package on planting and harvesting of nuts in practice at J&K also needs to be extended. At present, nuts, especially almonds, worth a few hundred crores of rupees are being imported into India from the USA and other countries. A considerable part of such import can be met indigenously or even the import itself substituted fully.

This book, in view of the above, is basically meant for the use of voluntary agencies, people's groups and training and extension institutes engaged in promotion and extension of nuts production, conservationists and environmentalists, farmers, educated unemployed, non-governmental organizations, personnel of the Agriculture Extension Department and others. The purpose is to create awareness among the people of the advantages of production of various kinds of nuts that have scope for development in the Himalayas. An attempt has been made to cover all important aspects of nuts production, including the present state and climatic requirements; cultural pre- and post-harvest practices; marketing and economics of production. It is hoped that the publication would be of great use to all concerned. A special Hindi edition will be brought out for the use of farmers of the Western and Central Himalayas. As the book also gives figures of production, imports and exports in selected countries of the world, it would be of good use for people in the field including the educated unemployed who could set up import-export business as self-employment.

TABLE 2.1: Prosperity through promoting the Himalayan Nut Production Programme

Food	Employment	Fuel	Animal Feed	Environment
* Providing energy-rich supplementary food to undernourished hill people. * Availability of fruits to the common people at low cost. * Improvement in the health of Himalayan rural population.	* Generating employment through engaging local inhabitants in nursery raising and other cultivation practices, harvesting, processing and marketing. * Checking migration of people from hills to plains. * Opening new avenues for unemployed youth,	* Providing fuel in the form of dead/diseased branches, pruned, branches, etc. * Reducing pressure on existing forest.	* Providing rich feed in the form of oilcakes for livestocks and hulls of almonds for hogs.	* Soil conservation. * Bio-aesthetic value. * Providing fresh air for good health and hygine. * Controlling different kinds of pollution: chemical, noise, dust and water. * Act as windbreaks.

(Contd.)

TABLE 2.1 *Contd.*

Food	Employment	Fuel	Animal Feed	Environment
* Promoting subsidiary nut-based industries in the region.	marginal and landless hill communities.			* Help in regulating hydrological cycle.
* Developing alternative food sources for the common masses.	* Engaging hill people in various nut-based industries.			* Improving soil fertility * Help in increasing forest area.
	* Improving the economic condition of local artisans by providing raw material for cottage industries, like wood-carving.			* Use of waste/degraded lands, community land for planting food-producing trees.
	* Providing raw materials for medicinal products like tooth powder, from walnut hulls and bark; apricot oil, almond oil for cosmetics, etc.			

The present work deals with the production of important commercial temperate nuts like, walnut, almond, chestnut, hazelnut, pecan nut, apricot, chilgoza, etc. The nut promotion programme in the Himalayan region is aimed at meeting the following broad objectives:

1. Creating awareness and motivating people for growing nuts on the degraded, waste and village community land.
2. Providing energy-rich supplementary food to the people of the region.
3. Rational use of land and human resource.
4. Improving environment and conserving important natural resources, viz. water and soil.
5. Control of soil and nutrient losses.
6. Involving the rural masses in various tree-planting, production and marketing activities.
7. Uplifting the socio-economic status of hill folk through employment generation.

Tabe 2.1 summarizes the elements of prosperity through the Himalayan Nuts Production Programme for food, employment, income-generation, fuel, animal feed and environment.

3

TREE NUTS OF THE WORLD

TREE NUTS are one of the oldest sources of food for man, bird and animal and their use in the form of food, branches and flowers has been in practice on religious occasions since antiquity as is evident from the Vedas and the Bible.

Walnuts (Juglans regia) were cultivated in King Soloman's garden; pistachio nuts (Pistacia vera) were carried down to Egypt by Joseph's brother; almonds (Prunus dulcis) were and are used in various pujas and religious ceremonies.

Many nuts are associated with the foods of forest dwellers of many parts of the world including the Himalayan region in India; also of the early Egyptians, Romans, Italians, Chinese, Japanese, Hawaiians, American Indians, Latin Americans, Mexicans, Africans and inhabitants of many parts of the Indian subcontinent.

In Table 3.1 are given the tree nuts of the world.

TABLE 3.1: Tree nuts of the world

Common name	Botanical name	Leading producing countries
Almond	Prunus dulcis Mill	Italy, United States, Iran, Spain, Morocco, Portugal.
Apricot nut	Prunus armeniaca L.	Europe, United States, China.
Beechnut	Fagus grandifolia	Denmark, Germany, United States.

Contd.

TABLE 3.1 *Contd.*

Common name	Botanical name	Leading producing countries
Betel nut	Areca catechu L.	South Sea Islands, Palau and Yap Islands in the Carolines.
Brazil nut	Bertholettia excelsa	Brazil, Bolivia, Peru.
Butter nut	Juglans cinerea L.	United States, Japan.
Cashew nut	Anacardium occidentale L.	India, Kenya, Mozambique Tanzania, Brazil.
Chestnut American	Castanea dentata Borkh	Eastern United States
Chestnut Chinese	Castanea mollissima Blume	China
Chestnut European	Castanea sativa Mill.	Europe
Chestnut Japanese	Castanea crenata Sieb and Zucc.	Japan
American chinquapin	Castanea pumila Mill.	Eastern United States.
Mamaron nut	Scheelea macronarpa	Brazil, Central America.
Filbert	Corylus avellana Linn. Coryclus americana	West Germany, Switzerland United Kingdon, France, United States, Czechoslovakia.
Japanese walnut	Juglans sieboldiana Maxim	Japan, United States
Hickory nut	Carya ovata (Mill) K. Koch. Carya glabra Carya laciniosa (Mich. F.) Loud.	United States
Macadamia nut	Macadamia integrifolia Maiden & Betche	Hawaii, Kauai, Maui, Oshu, Australia
Pecan	Carya illinoensis K. Koch.	United States, Mexico
Pine nut	Pinus spp.	Italy, Spain, Portugal, Iran, United States
Pistachio nut	Pistacia vera L.	Iran, Syria, Turkey, India, Afghanistan, United States
Black walnut	Juglans nigra L.	Central, Eastern and South-western United States
English or Persian walnut	Juglans regia L.	United States, France, Italy, India, Turkey, Iran, Yugoslavia
Chilgoza pine nut	Pinus gerardiana Wall	Afghanistan, Pakistan, India.

4

WALNUT
(Juglans regia L.)

WALNUT belongs to the genus *Juglans*. The only commercially important walnut is the English or Persian, *J. regia* L. which is grown in the north-west hill region of India. It is also known by the name of *Akhrot* in the greater part of the country. It is native to the north-western Himalayas and is grown in all parts of the Himalayan region between the elevation of 1,200 to 2,150m above mean sea level. In the hill regions of northern India, it is mostly grown as scattered trees and not in the form of compact and well-laid-out orchards. However, there is great scope for bringing more area under its cultivation in the form of regular plantation, especially keeping in view the demand from inside and outside the country, especially for the walnut kernel due to its high nutritive value. It contains 4.5 per cent moisture, 15.6 per cent protein, 64.5 per cent fat, 11.0 per cent carbohydrates, 1.8 per cent mineral, 0.11 per cent calcium, 4.8 per cent iron, 10 I.U. Vitamin B and 450 mg/100 gm Vitamin B. One hundred grams of walnut kernels provide 687 calories of energy.

Jammu and Kashmir is the principal walnut-producing State in India. It produces about 98 per cent of the nation's total crop. Except for some districts of Jammu, parts of Kathwa, Udhampur and Serz, walnuts are grown all over the

State. A few isolated pockets of walnut production are dispersed in the hilly areas of the neighbouring States of Himachal Pradesh and Uttar Pradesh. Apart from bringing in contributing foreign exchange, walnut production provides occupation to thousands of families engaged in shelling, lumbering, woodcarving and marketing of the fruit. During the last 10 years the walnut production has recorded between 15,000 and 20,000 metric tonnes per annum with an average annual production of 16,400 metric tonnes (during 1985-1989). The walnut season starts early in India and nut maturity commences towards the end of September. The flowering begins in March and the fruit setting and development takes about five to six months from blooming.

Most of the walnut area is rainfed; when available, farmers provide irrigation for 7-14 days before harvest. This helps the shucks to open and remain suspended on the trees, dropping clean nuts to the ground. The application of fertilizer is basically adhoc although the J&K Horticulture Department does issue guidelines to growers. But despite this and the fact that agro-climate conditions of the J&K State are ideally suited to walnut growth, limited progress towards varietal improvement has hampered increased production. However, the State is a rich source of walnut germplasm with about 330 useful strains isolated to date. The available strains offer tremendous genetic variability which can be further exploited by undertaking breeding programmes. Out of the total genetic base available presently in the State 65 strains exhibit high quality kernel and can rightly be used as mother stock.

Layout

Due to high population density and scattered small farm holdings, there are only a few compact orchards. Planting consists of a few dozen seedlings on farms occupying about 1 to 1.5 ha. Generally all important orchard operations are done manually. The growing sites represent favourable climatic conditions with good fertile alluvial soil which are ideally suited for its cultivation. The land almost levelled or with a moderate slope has been occupied by rice cultivation. The sites with more than eight per cent slope are being utilized for apple cultivation. Walnuts are grown in the areas located at an elevation of 1,800 m above mean sea level. The uppermost steep slopy ridges of the J&K hills have been put under forest. There are thousands of walnut trees planted in the forests for timber and wild nuts. A thriving wood-carving industry exists in the area.

Crop Area

Historically, walnut trees grow wild in the north-west Himalayan belt. But, in recent years, planting has become more systematized and structured. But, since the Ministry of Agriculture, Government of India, does not maintain any official data on the production or area of the walnut crop, the planted area in 1990-1991, based on the rough data provided by the J&K State Horticulture Department, other official and private sources, is reported to have increased marginally to 35,200 ha from 35,000 ha in 1989-90. The actual harvested area is estimated at about 70 per cent of the total planted area containing about 1.4 million trees, including the small orchards of the neighbouring

States of Himachal Pradesh and U.P. Some of the trees in the J&K State are several centuries old, having attained a height up to 45 metres and girths up to about 11 metres.

As can be seen from Table 4.1 there is no noticeable improvement in the status of walnut in India as forecast for 1991 compared to 1990.

(Table 4.1 show the walnut status in India 1991)

TABLE 4.1: Prodution Supply and Distribution of Walnuts in India (on shell basis)

Status of walnut	1990	1991
Area planted (ha.)	35000	35200
Area harvested (ha.)	24500	24640
Bearing trees (nos.)	868000	971000
Non-bearing trees (nos.)	410000	416000
Total trees (nos.)	1278000	1387000
Beginning stocks (m. tonnes)	980	480
Production (m. tonnes)	17000	20000
Total supply (m. tonnes)	17980	20480
Exports (m. tonnes)	8500	10000
Domestic consumption (m. tonnes)	9000	9800
Ending stocks (m. tonnes)	480	680
Total distribution (m. tonnes)	17980	20480

Yields

Due to long juvenility seedling trees start bearing at about 10-12 years of age. Vegetatively propagated trees particularly by patch budding, cleft and veneer grafting start early bearing at about 6-7 years of age. However, the per cent success rate has failed to meet the present nursery stock requirements of cultivators and, therefore, massive concentrated efforts by research organizations are needed to achieve a higher success rate in budded and grafted plants. Recently, the success rate in budded and grafted plants. Recently, the success rate in the case of budding and grafting has been 40 per cent and 25 per cent, respectively. The peak nut production period starts at the age of 17-20 years and optimum nut production duration continues for about 20 years. Walnuts produced in J&K have an average shelling percentage of 30-35 but can go as high as 65 in certain selections. Peak nut production is 80 kg, to 100 kg per tree. Over the productive life-span of the tree, estimated assuring average yield is between 45 kg and 50 kg.

Gross Commodity Development

In most of the orchards in J&K State crops such as mustard, rape, millet, vegetables and legumes are planted in between the walnut trees to prevent soil erosion, improve water retention and soil texture and also to enrich the soil by adding humus. The scattered nature of orchards makes it difficult to estimate a standard cost of production. Cost of in-shell walnuts under average bearing conditions is around Rs. 270 per 100 kg. Walnut production in the State ranks second in earnings after apples.

Type of Walnut

Based on the shell thickness trait generally three types of walnuts are grown in India. These are hard shelled, semi-hard shelled and thin or paper-shelled commonly known as kagazi. Kagazi is considered superior in quality due to its thin and easily cracking shell. Kernels obtained from this type of walnut are generally large in size and sweet in taste. Recovery of kernels from this shell walnut can be as high as 65 per cent.

Demand and Uses of Walnut

Demand for walnuts is on the increase, since they are cheaper than almonds and cashews they sell at about Rs. 25 per kg compared to Rs. 100 per kg for cashews and Rs. 150 per kg for almonds. Their usage in confectionary and icecream-making is increasing. They are also used in many other ways – raw, salted, bakery products and others. They are also used as a dry fruit and are highly priced as dessert during the winter season. In J&K State, about five per cent of the nuts are consumed in the raw form. Walnut oil, especially in the producing areas, is used for culinary purposes and about three to four per cent of the local production goes into oil production. Oil extraction rates vary from place to place from 45 to 65 per cent. Walnuts are also used as a colouring agent, extracting black and green dyes from the husk.

Price

The wholesale price of the 1989-90 season's in-shell walnut in Delhi ranged between Rs. 15 to Rs. 16 per kg. The wholesale price of ungraded walnut in Srinagar is around Rs. 12 per kg The wholesale price of walnut kernels (fair average quality) both in Delhi and Srinagar is around Rs. 60 per kg. Superior quality kernels are being traded at Rs. 65 per kg. The prices fall slightly after December, the peak consumption period.

Exports

Normally, total walnut production is split almost equally between the domestic and foreign markets. China, the main competitor of India, is selling its crop below the Indian prices.

Table 4.2 shows exports of walnuts from India during the period 1983-1988.

TABLE 4.2: Exports of Walnuts from India

S.No.	Year	Quantity exported in metric tonnes		Value (Rs. lakhs)
		In shell	Kernels	
1.	1983-84	449	3181	783
2.	1984-85	352	4630	1072
3.	1985-86	891	3518	820
4.	1986-87	300	4062	1152
5.	1987-88	256	3876	1256

The export of walnut from India was averaged at Rs. 1,016 lakhs per annum during the period 1983-88. It is regularly exported to over 20 countries, Europe being at the top. The European importing countries include France, the U.K., the Netherlands, Germany, Czechoslovakia and Denmark. Other importing countries include Jordan, Syria, Bahrain, the United Arab Emirates, Iraq, Australia, etc. Exports are more in kernel form due to their excellent quality resulting from manual shelling.

Imports

Import of walnuts is prohibited in India.

Export Subsidies

Apart from the continued cash subsidy of 10 per cent on f.o.b. value of walnut kernels, the Government of India allowed, in July, 1990, imports of dry fruits (under 10 per cent of f.o.b. value) against replenishment licences. The Government also allowed four per cent import replenishment to exporters on import of packing material (glassine paper). These incentives have been made available until March, 1993.

Promotion Policy

The J&K State Government continues to support the planting of walnut trees to conserve soil. Necessary inputs like planting material, fertilizers and pesticides are subsidised by the J&K Horticulture Department. The Government is also setting up laboratories for promotion of tissue cultures, multiplication of clonal plants and rejuvenation of old, inferior and less productive walnut orchards.

Quality Control

The Government of India has taken up many schemes to increase production as well as to improve the quality of walnuts. These include distribution of seedlings of superior varieties and introduction of modern devices for washing, sorting and cleaning. Research institutions are also engaged in experiments for developing new processing techniques. In order to ensure that quality walnuts are exported from India, a scheme for compulsory quality control and pre-inspection was introduced in 1963 and is still in operation. Jammu is the main grading and packing centre for walnuts; other centres are Delhi and Bombay. The Directorate of Marketing and Inspection, Government of India, examines the export consignments and issues an "Agmark" label at Jammu and in Bombay before the shipment leaves the country. ("Agmark" is the official mark of the Government of India given to agricultural and livestock products that conform to the grade standards laid down under the Agriculture Produce (Grading and Marketing Act, 1932).

The Walnut Grading and Marketing Rules, 1966 provide grade standards for both in-shell and shelled walnuts. Grades for in-shell walnuts are mainly used on the basis of size, whereas, for kernels, colour and size are the two main grade-determining factors. For in-shell walnuts, there are four regular grades and for kernels thirteen. Besides, there is an "X" grade for such quality of walnuts which cannot be graded under regular grades to meet the special requirements of foreign buyers.

The essential conditions for deciding the quality of in-shell and shelled walnuts are as follows:

In-shell Walnut

Walnuts of only the current year's crop can be graded. The nuts should be free from pests, grubs or eggs. They should be well developed, well washed or bleached and have a clean and attractive colour. They must also be reasonably dry so that the loss in weight does not exceed one per cent on arrival at destination. They should give over 90 per cent good cracking and yield edible kernels of acceptable taste and aroma. The lots have also to be full-grown nuts and should be free from defects like kernel dampening, oil seepage; mould or fungus attack, rancidity, shrivelling, insect-pest infestation and powdering. The tolerance limit for invisible or internal defects is 10 per cent of every count, out of which the limit for hard nuts is two per cent.

The nut lots brought for quality grading have to be fully developed and not be undersized, partially developed, deformed or damaged and should not have crack shell, split or perforations, oil stain, sunburn or blight mark, or residue of chemical bleaches. They have to be clean, well graded and fairly free from foreign matter like cob-webs, shell grits or hull remains.

The kernels of the current regular crop only are allowed to be graded. They should not include any diseased or mouldy pieces, and be free from cob-webs, rodent excreta, human hair, live pests, grubs or eggs, shell grits, wood splinters, husk and other foreign matter. Also, the kernels should have a pleasant taste and flavour. They must not be partially or wholly shrunken or shrivelled, oil-bled, darkened, blighted, sunburnt, worm-eaten, rancid, bitter, excessively oily, tainted or blemished. A tolerance of four per cent by weight is allowed for defective kernels.

Packaging and Labelling

Every graded consignment bears inner and outer Agmark labels on which grade, designation, date of packing and year of harvest are marked. Besides, the package will also indicate the place of packing. To ensure that walnuts are free from inherent or cross infestation, all consignments of shelled walnuts are compulsorily fumigated before shipment. For in-shell walnuts, fumigation is compulsory, if shipment is made between February 1 and September 30. At each centre of packing, an advisory panel is set up to settle quality disputes.

Sound clean gunny bags are used for packing in-shell walnuts. The conventional packing sizes are 25 kgs. and 50 kgs. Boxes made of well-seasoned wood or cardboard boxes with proper lining are used for packaging of shelled walnuts. The packaging sizes vary from 6.25 kgs to 25 kgs.

Walnut Cultivation Status in Himachal Pradesh

Compared to the State of Jammu and Kashmir, conditions for walnut growing in Himachal Pradesh are not so conducive. Consequently walnut plantations in Himachal Pradesh are not in a regular shape of orchards. They are mainly concentrated in the districts of Kinnaur, Sirmour and Chamba. Some plants are also seen in the districts of Shimla, Mandi and Kulu. One or two walnut trees can also been growing the hinterlands, marginal lands in villages between the elevation of 1,200 to 2,150 m above sea level.

It is estimated that in the whole of Himachal Pradesh there are about 5,000 trees. The total walnut production is estimated at only 250 tonnes which is mostly consumed in the State as a dry fruit. The quality of walnut is not very good and all the trees are raised from seedlings. Some good quality walnut trees are found in the cold and dry regions of Kinnaur district. A few selections of outstanding merit from chance walnut seedlings in Kinnaur district have been made. One of these is found to be having many desirable attributes and has been named as "Govind". The efforts at present are to propagate the plant material from this selection.

Major Constraints in the Cultivation of Walnut

The major problems coming in the way of successful cultivation of walnut in the north-west hill regions of India are:

(i) Lack of suitable methods of propagation and supply of adequate vegetatively propagated plant material,

(ii) Lack of suitable root-stock,
(iii) Lack of suitable superior strains/cultivars,
(iv) Ignorance about the pollination behaviour of the species and lack of suitable pollinizer coinciding with the blooming period,
(v) Lack of quality nursery stock which, in turn, affects the survival rate, and
(vi) Non-availability of improved package of practices.

Suggested Future Plan of Work

Keeping in view the above constraints a plan of work is suggested. This would include standardization of grafting and budding techniques; improvement in percentage success through various techniques of propagation, including use of growth regulators; working out of techniques of ripening of the seasoned wood before utilization in grafting and budding; suitable techniques for top working of inferior walnut trees; propagation of walnut plant through stem-cutting and stooling with the use of growth regulators under mist and glass-house conditions; new rootstocks such as *Juglans regia, J. hindsii, J. nigra, J. californica* and *J. repestris* and paradox hybrid walnut to be introduced; immediate introduction for adoption of outstanding marginal varieties from horticulturally advanced countries like the USA (Davis, California); comprehensive survey to study the existing improved strains; field experiments for finding out proper fertilizer requirement; study of floral biology of cultivars for selection of suitable pollinizers, etc.

Experience of Walnut Development in Jammu & Kashmir

The present position of walnut production and

related matters has already been discussed. Now, it is proposed to indicate in brief the future line of action. As already pointed out, the majority of trees growing in J&K are of seedling origin. These trees produce nuts of variable quality, attain a huge size and have a gestation period of about 15-20 years.

The older plantation is mainly growing in un-reclaimed areas and on poor type of soils generally on terrace walls, odd places or near dwellings. One of the most important method to improve walnut cultivation is to multiply and propagate vegetatively walnut plants of known high yielding commercial varieties/locally marked quality strains for plantation in regular orchards. In this context, the following steps deserve consideration :

(i) Sufficient number of planting material to imported from the USA for supply of following varieties:

1. Eureka
2. Placentia
3. Payne
4. Serr
5. Hartley
6. Tehama
7. Vina

(ii) There is no efficient, standard, commercialized technique for vegetative production of walnuts. However, machine-grafting for walnuts has been commercialized in the USSR. The machine is named M.P. 7-A and is in use at the Institute of Horticulture Kishenev, Maldevia, USSR. A few machines may be imported and supplied to the J&K and other hill States.

(iii) De-hulling is being done in the State manually. The process not only stains the hands of the workers but also injures the hands due to the high toxic acid content of the processing agents used. This warrants use of small mechanical hullers which are in use in other walnut-producing countries and are more efficient. Likewise, after hulling, nuts are washed in J&K in the naturally flowing streams. This is rather inefficient method. Drying is usually done under the sun. Because of low temperature, the process is prolonged and nuts are sent to Jammu in half-dried condition. These inefficient methods of post-harvest processing render almost one-third of the total produce unfit for the export market.

To sum up, use of mechanical hullers, efficient method of cleaning and bleaching of nuts and use of dehydrators needs to be encouraged. Three small machines have been imported from Italy under an Indo-Italian Project and are being demonstrated and put to use (under the project) in the walnut-growing areas of the State. Two drying machines have also been imported from Italy under the Project for demonstration purposes. The number of small mechanical hullers and drying machines needs to be increased. These steps, if taken, would increase the exportability of walnuts under international market standard requirements.

World production of walnuts at a glance

It may be worthwhile to give a production picture of walnuts in the major producing countries. Table 4.3 shows the position for the period 1980-88.

TABLE 4.3: Walnut production in major producing countries (1980-88)

Country	1980	1981	1982	1983	1984	1985	1986	1987	1988
France	18000	7000	35000	16000	14000	30000	20000	22000	13000
Italy	18000	18000	15000	15000	15000	15000	10000	25000	15000
Turkey	6000	10000	12000	10000	5500	10000	11000	5000	5000
USA	177800	204100	211400	163300	190500	195000	169600	222300	185475
India	18000	17000	20000	20000	15000	15000	18000	16000	16000

It is clear from the above Table that the USA leads in the production of walnut: its share is about 79 per cent.

Major Walnut Importing Countries

Europe, Africa, the USA, Canada, Asia and Oceania are the major importers of walnuts (see Table 4.4).

It will be seen from Table 4.4 that the largest importer of walnuts is West Germany, followed by Canada, the U.K. and Japan.

Major Walnut Exporting Countries

The USA, China, France, Italy and Turkey are the major walnut exporting countries (See Table 4.5).

It can be seen from Table 4.5 that the USA is the largest exporter of in-shell walnut, followed by China and France. In kernel, China is No. 1 exporter followed by the USA and France.

Improved Agro-Technology for Quality Walnut Production

Yield and quality of fruits are significantly affected by several contributing factors, which are closely associated with the crop management technology followed for a crop. Based on the experience of the progressive walnut growers throughout the country and abroad and the conclusions drawn from the experimental findings, the improved agro-technology to manage the crop for high profitability is being given below.

TABLE 4.4: Major Walnut Importing Countries (figures for 1984-88)
(M. Tonnes)

Country	1984	1985	1986	1987	1988
Europe					
Austria	1295	1968	2028	2241	2157
Belgium	1942	1796	1335	1863	1900
Denmark	918	925	962	1029	1090
Finland	28	—	—	—	—
West Germany					
(in-shell)	19335	18132	20006	20102	17671
(kernel)	1674	1699	2086	2349	2307
Netherlands	2585	3058	2798	3505	3220
Norway	517	520	510	582	593
Sweden	452	444	463	505	529
Switzerland	1836	1967	2035	2079	2189
U.K. (in-shell)	2826	2432	3173	3733	2404
(kernel)	3854	4115	4211	4453	4000
Africa					
South Africa	141	84	—	—	—
America					
Canada					
(in-shell)	2578	2254	2205	1852	2539
(kernel)	4794	5548	3849	5855	5107
USA 45	149	124	1233	139	
Asia and Oceania					
Australia					
(in-shell)	1695	678	753	645	518
(kernel)	1942	1709	2840	2042	1829
Japan	2396	2101	3881	6401	4426
New Zealand	240	266	230	260	469

Note: Unless otherwise specified figures include in-shell and kernel.
— Not available

Climatic Requirement

Walnut plant survives well in those places where the

autumn is free from frost because the season coincides with the blossoming period and would cause wilting of newly born flowers. The normal development of the fruit requires mild sunlight in summers and an average rainfall of between 50 and 70 cm is favourable for its cultivation.

TABLE 4.5: Major walnut exporting countries (figures for 1984-88)
(M. Tonnes)

Country		1984	1985	1986	1987	1988
USA	(in-shell)	39530	44045	45038	52997	48219
	(kernel)	6190	7423	8528	7375	9377
China	(in-shell)	11244	11844	13786	9777	8320
	(kernel)	5563	8212	8212	11293	10608
France	(in-shell)	4014	4286	4582	5480	4286
	(kernel)	1819	1913	2086	2331	2659
Italy	(in-shell)	1245	1512	766	360	1033
	(kernel)	1062	262	281	680	580
Turkey	(in-shell)	159	300	20	350	200
	(kernel)	1366	1200	1350	550	400

Soil Requirement

The cultivation of walnut can be done in any soil but silty loam has proved better for proper growth and development of the plant.

Propagation

Generaly, the propagation of walnut is done by seed or grafting (vegetative).

(a) *Seed Propagation*

Healthy and disease-free seeds should be selected for sowing. They may be sown in lines 50 cm apart and the seed-to-seed distance be kept at 25 cm. Sowing is done during mid-November to mid-January depending on the altitude and temperature. Higher germination rates in walnut have been achieved by sowing seeds in a vertical position at a depth of 7 cm (Lal, *et. al*, 1984).

The beds after sowing and irrigation should be covered with grasses. While preparing the beds, proper drainage facilities should be provided. However, it is advisable to avoid propagation by seed because the plants which are grown by this method take a very long time (10 to 12 years approx.) in fruiting and the plant's characteristics may not be true to type.

(b) *Vegetative Propagation*

Tongue or whip-grafting and patch-budding are generally practised to propagate walnut plants vegetatively. The best period for grafting is January-February, and for budding May-June. For propagating the plants through veneer grafting, 5-6 month-old scion wood of pencil thickness having a length of 15cm is grafted on to the rootstock of same thickness. The selected scion wood should be defoliated 15 days prior to its detachment from the scion cultivars. The optimum time for veneer grafting under mid-hill conditions is July-August. One-year-old seedling of hard-shelled walnut or black walnut (*J. nigra*) can be used as rootstock. Scion for patch-budding or tongue-grafting should always be selected from the tree which has already started fruiting.

Planting Time and Plan

Pits of the size of 1.25×1.25 m at a distance of 10×10

m should be dug during September and should be filled up with orchard soil, 50 kg well-rotted FYM, 150 gm aldrin dust, 150 gm urea, 500 gm each of superphosphate and muriate of potash, all thoroughly mixed. The best season for walnut plantation is December to January. The plant should be well fixed in the soil and the adjoining portion of scion and stock union should be at least 15 cm above the ground surface. Watering should be done soon after plantation. To protect the plant from collar rot it should always be treated with Dithane Z-78 before planting.

Training and Pruning

Generally walnut requires very light pruning but to frame the plant in a good shape branch-pruning should be avoided up to a height of one metre from the main trunk of the plant. But the dried and diseased branches should be properly cut off from time to time.

Manuring

Walnut tree has a vigorous growth and as such requires heavy feeding. For the proper vegetative and reproductive growth, a balanced nutrition of nitrogen, phosphorus and potassium should, therefore, be applied at proper time. A one-year-old seedling requires 500 gm urea and 750 gm each of superphosphate and muriate of potash. The doses of fertilizer may be increased with the increase in the age of a tree. The full dose of superphosphate and muriate of potash and a half dose of urea should be applied in October after picking the fruits and the remaining half dose in the month of February every year. Whenever fertilizers are applied these should as said above, be well mixed with the soil and should always be applied at least 30 cm away from the main shoot of plant. Every year in the month of October, 50 kg of well-rotted FYM should also be applied as a normal feed. To check the physiological disorders in the plant due to any deficiency of micronutrients, foliar application of all trace elements like zinc, copper, manganese, boron, iron and magnesium may be done in the month of March.

Time of Flower and Fruit Setting

Walnut trees start flowering from April to May. The initiation of both male and female flowers takes place separately. The setting of fruits starts from the last week of May to June and maturing of fruits is over by September/October.

Diseases and Insect-Pest Management of Walnut

There are various walnut diseases. The symptoms and control measures are described below:

(i) *Apical Decline*

In this disease the apical portion of the young twigs starts declining and gradually the whole of the plant dries up. To protect the plant against this, spray of Dithane Z-78 or Dithane M-45 should be done at fortnight intervals. The diseased twigs should be trimmed off and burnt.

(ii) *Leaf Spot*

It is a fungal disease and the first sign of its appearance are brownish spots visible on the upper layer of the leaves and grey coloured spots on the lower surface of the leaves which afterwards also change to brownish spots. In the later stage, the disease may cause complete drying up of a plant. To avoid this disease, dried and fallen leaves should be burnt and a spray of fungicides like Dithane Z-78 or Dithane M-45 should be done in the month of February every year.

(iii) *Chestnut Blight*

The causal organism of Chestnut Blight is a fungus known as *Indothia parasitica* which becomes active during April-May and causes decline of new twigs and foliage. Splitting of old branches also takes place due to it and there appear brownish spots on the bark of the twigs. In the severe stage of infection, the bark of the tree is detached from the twigs and the stem. A spray of Bordeaux mixture of 5:5:50 may be done to protect the plants from this disease.

Insect-Pests

The walnut plants are generally infested by root and stem borers and tent, fruit and leaf-eating caterpillars. To protect the plant against root borers, application of 225 gm of 5 per cent Aldrin dust should be done 25 cm deep in the soil during June-July, while a spray of 0.03 per cent metasystox can protect the plants from stem borers, tent, fruit and leaf-eating caterpillars. If there is an attack by stem borers and there appear holes on the stem, cotton gauz wetted with a mixture of 2:1 chloroform and creasote oil should be kept in the holes. After keeping the gauz, holes should be closed with the help of clay soil.

Cleaning, Drying and Grading

After picking fruits from the trees, the diseased fruits should be isolated and healthy ones washed in water and dried in shady but open place. The drying of fruits may also be done in mild sunlight or by a dehydrating machine at 44.5° C temperature. Grading of fruits may be done according to their size and quality.

Yield

Generally a normal full-grown healthy tree of walnut yields about 40-50 kg of nuts annually.

Promotion of Walnut Production in H.P. and U.P. Himalayas under Trees for Life

The Trees For Life is promoting nut production in the Central Himalayas under a joint project, TFL/Tata Tea Ltd., with a base at Ranichauri. More recently, under the Trees For Life, the Dalai Lama Project is distributing carefully selected walnut seeds from Banihal (J&K) to about 400,000 devotees of the Dalai Lama and others in Dharamsala, Kangra district and other parts of the H.P. Himalayas. A small pamphlet (reproduced below) is also given to all receiving the "blessed" seeds.

Tips to Remember

When to grow ?

- The best time for sowing walnut seeds is December to January.

Where to grow ?

- In hills, between altitudes of 1,200 and 2,150 m above mean sea level.

How to grow ?

Step One – Preparation of Pits

- Dig pits of 1×1 m size. Empty the pit completely. Leave the pit exposed for 15 to 20 days,
- Ensure a spacing distance of 10×10 m between pits,
- Prepare the soil mixture by mixing one part soil with one part of well-composted farmyard manure. Add generous amounts of neem leaves or neem cake to prevent termites,
- Mix well, remove stones, pebbles, etc.,
- Fill the pits with soil and manure mixture, and
- Pour water to allow the soil to settle down.

Step Two – Sowing of Seeds

- Coat the seeds with red lead oxide mixed with linseed oil in 1:1 ratio. This is for rodent control as rats are able to take the uncoated Walnut seeds more readily,
- Place two seeds in each pit, intact with shell, with the apex side of the seed, vertical. Cover the hole with soil mixture,
- Ensure that when sown, only 7 cm of soil should be put over the seed shell,
- Press the soil over the seed hole to avoid air gaps,
- The ideal time for sowing is during winters. In

case there is no snowfall, irrigate with very little water, only at the time of sowing. Do not disturb the pit for two or three months, until the seed germinates,

- The seeds will sprout by March-April. Ensure normal maintenance of the seedlings. Transplant one of the seedlings in another pit after four months, and
- In order to avoid the accumulation of toxic substances exudated through roots and barks of *Juglans* species, the growers are advised to change the nursery site every year, otherwise there is every likelihood to get the allelopathic chemical, i.e. Juglone (Polyphenolic naphtha quinone) accumulated, which adversely affects the soil conditions at later stages and does not permit to grow other surroundings vegetation (Noggle and Fritz, 1986).

Maintenance

- Look after the tree till it is fully grown. Consult your local horticultural experts if you have any problem.

Fruiting

- The trees will start bearing fruits after about 10 years. You can expect a commercial crop after 12 to 15 years.
- Fully grown trees after 15 years normally yields about 20 kg increasing up to 70-100 kg annually after about 20 years.

Why Grow Walnut Trees ?

- Walnut trees have a life span of over 200

years,
- Walnut is one of the richest sources of energy. The nut is also used in a veriety of sweets and dishes,
- The roots of the tree help in soil conservation,
- After 20 years, your tree will give you additional income from over 70-100 kg of walnuts year after year for over 180 years, and
- The wood of a walnut tree, when it is cut after its full life-span is over, is of very high commercial value.

Special Request

- These seeds have been specially blessed by His Holiness the Dalai Lama. Please sow them as per instructions to get the best results,
- The walnut seeds are of the highest quality and have been procured with great difficulty and at a very high cost. Therefore, please do not waste the seeds, and
- If you do not have the space, please gift the seeds or saplings to people who have the space and are genuinely interested. They will plant them with reverence and bless you for the special and wonderful Tree for Life gift.

The Himalayan Nuts Promotion Programme has distributed about two tonnes of walnut seeds to the following six voluntary agencies which are involving several nurseries as well as about 1,000 farmers in the programme over the next two-three years.

1. Uttarakhand Jan Jagriti Sansthan Khadi, Jajal
   (Shri Pratap Shekhar).

2. Parvateeya Jan Kalyan Samiti Ranichauri (Shri Yogesh Chandra Bahuguna).
3. Sabli Mahila Mandal (Savitri Bahen, President).
4. Berinag Gram Swarajya Mandal Kandey, Pithoragarh, Kumaon (Shri Sadan Mishra, Secretary).
5. Central Himalayan Rural Action Group (CHIRAG) (Shri Kanai Lal, Chairman).
6. Society of Services – Prakriti (Col R.N. Khanna).

In addition, almonds and apricot seeds from J&K are being distributed but in lesser quantity. Sweet – chestnut seedlings raised at Berinag Gram Swarajya Mandal nurseries in Pithoragarh district are being distributed to all the five voluntary agencies.

A training programme for proper planting and caring was conducted in the winter of 1990 at Ranichauri. The trainers have been –

(i) Technical Experts of Hill Campus (G.B. Pant University of Agriculture & Technology), Ranichauri, who have also planted several quintals of walnuts in their horticulture nursery for distribution to the interested farmers in 1991-92.
(ii) Shri Sona Ullah Banihali, (Award Winner)
(iii) Shri Avinash Bakhri.
(iv) A Planter volunteer with at least five years' experience of planting walnuts, etc., in Banihal and other parts of J&K.

PLATE 1: Male and female inflorescence, germination stages and genetic variability in walnut.

A. A bunch of staminate (male) flowers.

B. Pistillate (female) flowers with exposed stigma.

C. Different stages of germination (a to f) showing opening of suture (a), emergence of radicle (b), development of radicle (c), development of secondary roots (d), three-month-old seedling (e), and one-year-old seedling with well-developed root system ready for outplanting (f).

PLATE 2: Demonstrating walnut sowing methods under the Himalayan nut production programme.

PLATE 3: Fruiting behaviour and genetic variability in Almond

A. A bunch of almond containing hull intact fruits

B. Variability in shape and size of nuts and kernels øf some promising cultivars

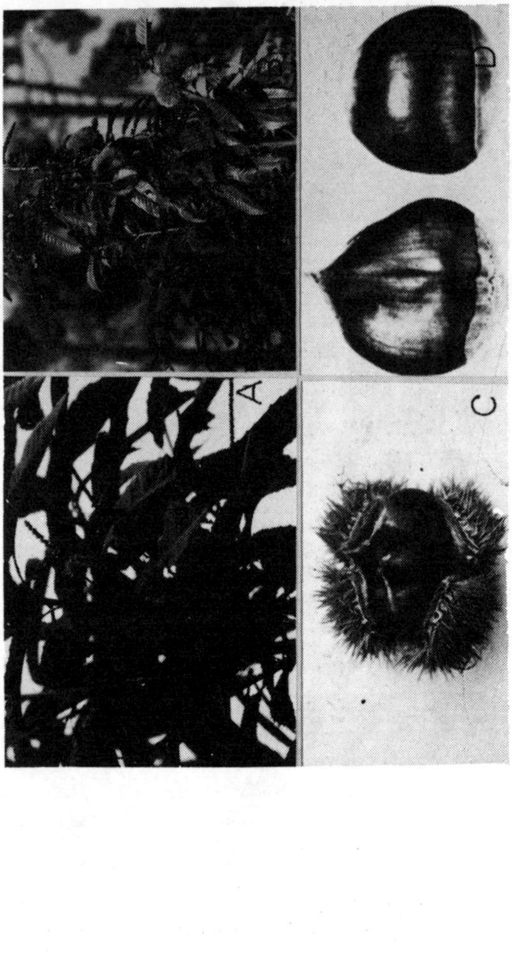

PLATE 4: Pistillate and staminate flowers, fruiting behaviour, opening of bur and mature dehulled fruits of chestnut

A. Twigs of chestnut containing staminate (calkins) and thorny pistillate flowers.

B. Eight-year-old tree containing fruits with intact bur.

C. Opening of bur and nut is ready to fall out.

D. Nuts of Japanese chestnut.

PLATE 5: **Showing staminate, pistillate, newly set fruits and mature hazelnut**

A. Staminate flower (catkin)

B. A cluster of young pistillate flowers.

C. A branch showing newly set fruit.

D. Newly developed hazel fruits along with mature nut.

PLATE 6: Depicting staminate inflorescence, fruit set half-grown stage, kernels and in-shelled fruits in pecan nut.

A. A branch containing staminate flowers.
B. In-shelled fruits of half-grown stage.
C. Compact kernels of pecan Cv. Desirable.
D. Kernels and mature fruit Cv. Stuart.

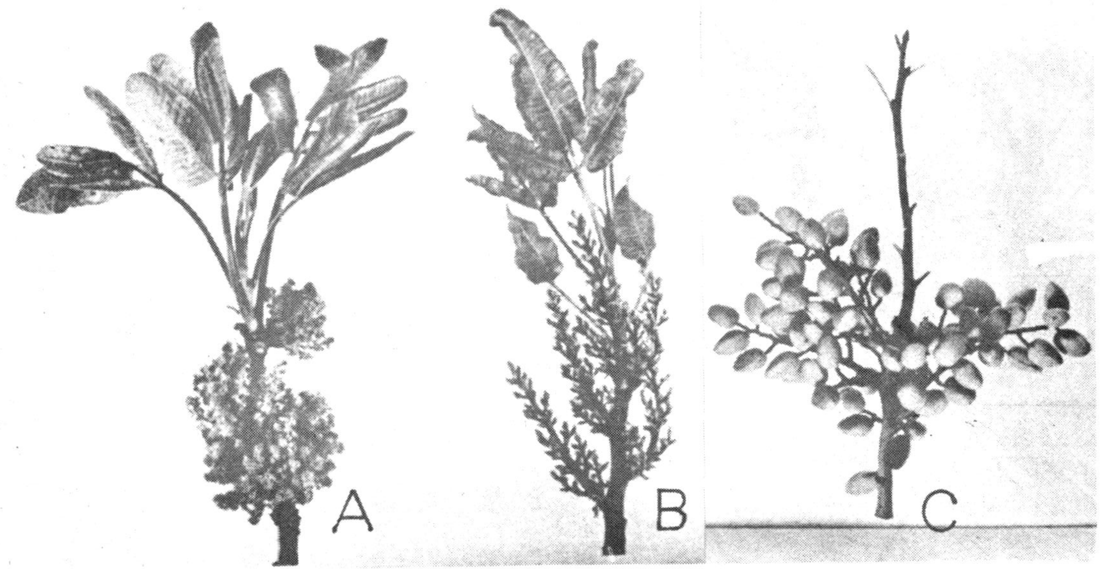

PLATE 7: Pistachio inflorescence and fruits.
A. Staminate flowers.
B. Pistillate flowers.
C. Nut-bearing branch.

I

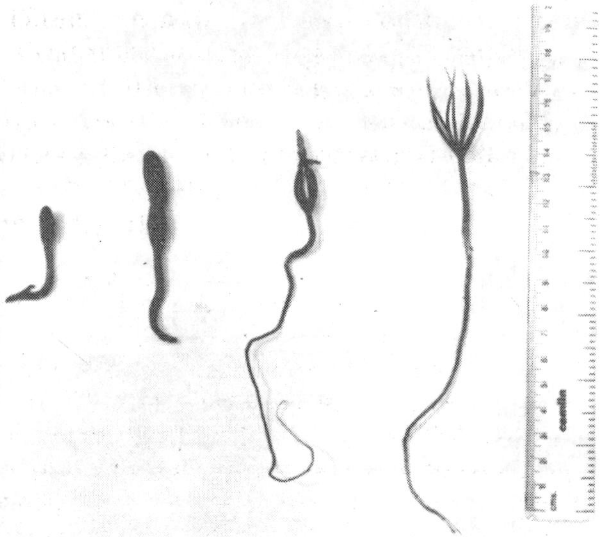

A | B | C | D

II

PLATE 8: Chilgoza pine nut and germination trend.

(I) Chilgoza pine nut.

(II) A. Emergence of radicle.

B. Development of cotyledons and radicle.

C. Seed coat ready to fall off.

D. Newly emerged seedling showing cotyledons and well developed root

PLATE 9: Fruiting behaviour and fruits of Apricot.

A. A limb containing Apricot fruits

B. Fruits of Apricot Cv. Turkey showing edible nuts and fleshy pulp.

PLATE 10: A. Showing cluster of Macadamia nuts

PLATE 11: Almonds.

PLATE 12: **Walnut Nursery.**

5

ALMOND
(Prunus dulcis Mill)

ALMOND is a native of the Mediterranean region. On account of the high quality and value of this nut, its cultivation has been tried in almost all the agricultural regions of the world. But, because of its specific climatic requirements, the attempts at most of the places failed and it is restricted to a few selected areas only. The leading almond-producing countries are Italy, the USA, Spain, Morocco, Portugal, Turkey, France, Iran, Algeria, Tunisia and Afghanistan.

In India, almond is the most favoured nut with the people regarding it as highly nourishing and of great medicinal value. A very valuable oil called *Badam Roghan* is also extracted from it. The major almond-producing area is Kashmir. But, recently, the possibility of growing this nut successfully in the cool and dry areas of Himachal Pradesh especially the Chini area bordering Tibet and to some extent in Lahaul has been realized. A small number of almond trees planted in these areas have given good results.

Unfruitfulness and its Causes

Three causes of failure in cultivation mainly responsible for unfruitfulness and posing threat of great losses to the almond orchardists are, firstly, self-incompatibility in most of the cultivars and trees shows unfruitfulness for want of suitable pollinizers due to different blooming periods of the planted cultivars. Secondly, the cultivars blooming earlier particularly in early spring do not bear any fruit due to freezing injury of bloosoms. Thirdly, the trees exhibit unfruitfulness if subjected to injuries caused by diseases and insect-pests and also due to deficiencies of some major and minor nutrients.

Climatic Requirement

Almond requires a cold and dry climate but fairly warm weather during the ripening period. Hilly areas of medium and high altitudes between 1,500 and 2,500 m above mean sea level with a rainfall of 50 to 60 cm are suitable for its cultivation. For proper growth of the almond tree and the nut production, the temperature should be less than 7°C for at least 60-80 days during winter. The almond plant flowers very early in spring. It needs dry and warm climate during this period. Frost and hailstorm are very injurious for almond growth.

Soil Requirements

The almond tree is deep-rooted and heavily draws all the essential nutrients from the soil. It gives heaviest yields when grown in deep, fertile and well-drained soils with a pH range between 5.8 and 6.5. Light loam soils are considered suitable for the almond. Very heavy soils are not at all suitable.

Propagation

In India, the common practice is to raise the plant from the seed of thin-shelled varieties and then plant the seedlings. Planting of seedling is considered to be a satisfactory practice, because each seedling tree being a separate variety, the pollination problem which is rather serious in the case of almond is solved.

For raising almond plants vegetatively, the methods of ring-budding in April-May and "T" budding in June-August are employed. Grafting is also done during winters by the cleft or tongue methods. The root-stocks commonly used are apricot, bitter almond and almond seedlings. However, bitter almond root-stock is better than others.

Varieties

As the almond trees growing in different States in India are of the seedling origin, there are no named standard varieties in existence. However, some exotic varieties have been introduced from the USA and have been tried in Kashmir, Punjab (low chilling cultivars), Himachal Pradesh and Uttar Pradesh. A brief description of the important varieties is given below :

Non-Pareil

This is one of the most important variety of almonds. It is paper-shelled and bears regularly. It commands a high price. The trees come into bearing early. The nut is elongated, flattened and pointed at the apical end. The shell is white and cracks with the slightest pressure of fingers. The kernel is flat, light brown, sweet with good flavour.

I.X.L

This variety gives good yield in the foothills and

valley areas. It is a desirable variety from the market stand point. The variety is incompatible with non-pareil and should not be planted with the latter.

Ne-Plus Ultra

This variety is good pollenizer for non-pareil, The nut is big and the shell is attractive. This enhances its market value.

Drake

This variety has a marked tendency to alternate bearing. It yields medium-sized nuts of good shape and colour. It is also a soft-shelled variety. The nut is oblong with rounded tip or slightly tapering. The shell is whitish or light brown in colour. The kernel is brown, elongated and well developed and sweet and delicious in taste.

Thin-shelled

This variety holds good promise of success. It is early and regular bearer. The nut is light brown in colour, attractive and smooth. The shell is very thin, rather papery. The kernel is brown, well developed, sweet and delicious.

Planting

The saplings are planted 6.5 – 8.5 m apart in later winters. As the almond varieties are mostly self-incompatable more than one variety should be planted in a block. Ordinarily, one variety will pollinate another variety, provided their flowering periods coincide with each other. It is also a good practice to plant in one seedling orchard. By this, the difficulties of self-incompatability of the grafted varieties will be overcome. A good arrangement is to plant the major portion of the area with a leading variety such as Non-pareil; every third tree in the third row may be of the pollinating variety.

Before plantation, pits of 1×1×1 m at a distance of five metres should be dug. Pits should be filled with a soil mixture containing 30 kg of well-decomposed Farm Yard Manure and 150 gm of BHC (5 per cent). The plants should be watered immediately after planting. Saplings should be planted in between December and February. These should be watered at least two-three times during summers.

These periodic irrigations during dry summers help in the proper development of the nut and prevention of fruitdrop and in laying the foundation of the next year's crop. In Kashmir, almond is grown under rainfed conditions.

Pruning

The fruit is borne largely on short spurs and to a small extent on the one-year-old wood. The spur remains productive up to five years and, therefore, one-fifth of the old bearing wood in the bearing trees should make an annual extension growth of 22.5 cm to 45 cm and in older trees from 20 cm to 30 cm. Pruning mainly consists of thinning out of surplus branches and cutting of diseased ones that hamper the growth of the plant. "Chaubatiya Paste" should be applied on the cut portion. However, a normal almond plant does not need much pruning.

Manuring

After planting, manure of the following composition is recommended per tree per year depending on the life of the plant:

Compost	10 kg
Nitrogen	30 gms
Phosphorus	20 gms
Potassium	20 gms

The above quantity should be increased every year up to six years depending upon the life of the plant. The application of fertilizer should be done twice a year first in December with half the quantity of nitrogenous fertilizer along with full quantities of compost, phosphorus and potassium and the remaining half of the nitrogenous fertilizer applied during August-September.

Harvesting

The fruit gets ready for harvesting in July to September. At this time, the hull splits and exposes the nut. Harvesting should be started when the hulls split and before the nuts begin to fall. A sheet of cloth is spread underneath the tree and the nuts are made to fall by giving the branches a vigorous shaking. The nuts are hulled, i.e. cleaned of any remaining dried skin. They are then sundried for a short period. Over exposure to sun should be avoided as, otherwise, the shells turn black and become unattractive. The nuts, especially of the hard-shelled type, are further bleached by sulphur fumes which make the shell bright and increase the market value of the nuts.

Yield

Almond trees begin to produce light crops in three to four years and reach optimum production in 8-10 years. A yield of 800 to 1130 kg per ha is common in many countries. Kernel recovery from the whole nut, on an average, is from 35 to 45 per cent.

Pest and Disease

The almond tree should be sprayed with a diesel oil emulsion in early stages of its growth because it is at this time that the Sanjose scale attacks it. But as the tree grows older, it becomes less susceptible to this pest. Leaf-eating beetles also attack the almond tree and defoliate it during summer. They should be controlled by using light traps and spraying of lead arsenate. The leaf-curling aphid is also a serious pest of almond at certain places. But it can be checked by spraying the tree with modern systemic insecticides. The trunks of the old trees are generally affected with gummosis. To overcome this problem the trunks should be painted with Bordeaux paste up to a height of 60 cm from the ground level.

Uses

Almond is a first-rate nut and used as a dry fruit. It is greatly used in confectionery. In tropical and sub-tropical regions, the nut is used for preparing delicious and nutritious cold drinks during the summer. The green almond is regarded as a great table delicacy. As mentioned earlier, it gives a valuable oil *(Badam Roghan)* which is of great nutritional and medicinal and industrial value.

World Almond Crop

The USA tops in production of almonds, followed by Spain. India imports almonds both in shell and kernel worth about Rs. 10 crores per annum mostly from the USA, Afghanistan and Iran.

Table 5.1 shows the world almond crop by major countries during the period 1984-88.

TABLE 5.1: World almond crop

(M tonnes kernels)

Country	1984	1985	1986	1987	1988
Greece	15000	16000	15000	8000	18000
Italy	12000	12000	14000	11000	12000
Spain	43000	65000	52000	70000	40000
Portugal	4000	3000	3000	3000	2000
USA	266200	210800	111000	297000	266000
Others	18000	18000	19000	19000	20000
Total	358200	324800	214000	408000	358000

Source : Woodroof, J.S. (1979), Tree Nuts-Production, Processing Products, AVI Publishing Co., inc. Westport, Conn, USA.

TABLE 5.2: Export of almonds (1984-88)

(M Tonnes)

Country	1984	1985	1986	1987	1988
France	909	774	2020	1226	649
Greece					
(in-shell)	*	*	*	*	*
(kernel)	2013	3195	2989	1632	—
Iran					
(in-shell)	—	—	—	—	—
(kernel)	—	—	—	—	—
Italy					
(in-shell)	10990	2771	7501	2400	3310
(kernel)					
Portugal					
(in-shell)	570	700	900	398	330
(kernel)	3799	1000	1300	800	325
Spain					
(in-shell)	1030	1841	1131	1210	1223
(kernel)	25172	21084	25064	25458	11874
USA					
(in-shell)	2669	6073	3624	4770	12533
(kernel)	79126	129447	100894	92844	141405

Notes: (—) Figures not available.
* Less than 500 kg.

Export of Almonds

The USA tops the export followed by Iran whose latest figures are not available. Spain is also an exporter. Afghanistan though not offering in this table is an exporter of almonds to India and Pakistan. Table 5.2 shows the World position.

6

CHESTNUT
(Castanea sativa Mill.)

CHESTNUT is a close relative of oak (*Quercus*) both belonging to the Fagaceae family of the cup-bearing trees. The nut of the chestnut is morphologically similar to the acorn of the oak. The chestnut bur is a vegetative structure comparable to the acorn cup. The beech (*Fagus*) also belongs the same family as chestnut and oak.

While chestnut trees have grown in most temperate climate of the world for more than 4,000 years for beauty, fuel and shelter, the food value of the nut has always been of prime importance. Productivity, ease of harvesting, simplicity of preparation for eating and diversity of uses for high and low-income people have been the impelling urge for growing of chestnut.

The marketing outlook is one of the chief considerations in evaluating a chestnut orchard. Three aspects of a successful marketing programme may be listed as follows :

(i) Development of early maturing cultivars to get a lead over other varieties;

(ii) Production of quality product at a moderate price (this includes grading and storing the nuts to assure a uniform product); and

(iii) Marketing the shelled nuts as clean kernels in polythene bags, both as fresh and frozen products.

The American chestnut provides food for a number of wildlife species also. Loss of chestnut in some areas has resulted in a great reduction in many species and complete disappearane of some as in the case of wild turkey in the USA. Recognizing the value of chestnut as food for wildlife, officials in several States in the USA began planting Chinese chestnuts on gap preserves 30-40 years ago.

For many years, the U.S. Department of Agriculture and the Connecticut Agricultural Experiment Station conducted research to develop a timber-producing tree by crossing the Chinese chestnut (*Castanea mollissima* Blume) with the American species, (*C. dentata*). It was found that to develop a timber tree for reforestation purposes, it would be necessary to in-breed first-generation hybrids until uniform progeny of the desired type resulted.

Diller *et al* (1964) has been the foremost proponent of reforestation with the Chinese chestnut and for many years he planted Chinese chestnut seedlings in the State and national forests in east USA with a standard spacing. He concluded that large plantings in timber are not recommended because the Chinese chestnut required good fertile land for maximum timber production with closely spaced trees.

Climatic Requirements

Chestnut requires a cool climate and flourishes in the mid-hills at elevations ranging from 1,500 to 2,100m above sea level. In India, chestnut trees are found in Kashmir, Himachal Pradesh, Kulu valley and Shimla hills, and Kumaon hills in Uttar

Pradesh.

Soil Requirement

Chestnut requires a deep fertile loamy soil with good moisture content. Experience has shown that the Chinese chestnut grows best on a sandy or sandy loam soil that is well drained. As a rule, the soil should be moderately to slightly acidic. The trees will not thrive well on a low ground that is poorly drained.

As much attention should be given to selecting a site for the Chinese chestnut as for the soil in which they are planted. Such trees should be located on a high sloping ground in order to provide adequate air drainage to prevent damage from frost.

In choosing a site for planting the Chinese chestnut, it is well to plant a few trees on a trial basis until it is determined that the site is suitable for orchard. Until more information is available on the cultural practices, and the production of the new crop, it is best not to devote high-priced land to a new crop.

Propagation

Chestnut trees in India are invariably propagated through seed. Chestnut seeds are planted some time after harvest (a storage treatment of one to two months at 0° to 2.2°C is necessary to the dormancy of chestnut embryos and to ensure uniform germination).

The nuts are usually planted 15 cm apart in rows 90 cm wide to allow for intensive cultivation. Seeds are planted 5 cm to 7.5 cm deep or three times the diameter of the nut. The rows are cultivated often during the first year of growth and all weeds are removed to give the nut seedlings proper spacing to grow and to eliminate competition for moisture and nutrients. When the seedlings are 15 to 20 cm tall, fertilizer may be applied by one handful of 5-10-5 of mixed fertilizer to each row of 1.2 metres. The fertilizer should be placed in a band on each side of the row at a distance of 15 cm from the seedlings and well mixed into the soil.

Special methods of propagating chestnuts include the buried-in-arch developed by Jaynes (1961). He (1964) also developed a method of rooting cuttings by use of sprouted seeds as nurse plants. At present, these methods are largely used in research, but with further refinement they may find more extended use in rapid nursery raising.

Budding of chestnut is not widely practised because one wood is fluted or groomed and the combination of bud and stock does not join uniformaly.

Three main essentials to achieve higher success in grafting are (i) good scion wood, (ii) proper training, and (iii) suitable technique.

Nursery-grafting is practised for propagating small trees for transplanting. Splice-grafting using scion and stock of the same diameter is the most satisfactory technique. Whip-grafting, a modified form of the splice technique, is also in use.

Two important considerations in nursery grafting are (i) to allow the rootstock to leaf out fully and (ii) to use dormant scion wood. The buds of scion wood must be fully dormant so that they do not dry out after grafting. Dormant buds are provided with protective bud scales to prevent water loss. When buds become active and start growing this protective covering is lost.

Inlay Bark Graft

A large chestnut tree has two objectives : (i) quick sprouting of the new crop and (ii) improving a poor producer. The inlay bark graft is the most satisfactory.

Cultural Practices

No special cultural treatment is given to this tree. Once established it flourishes without much care. However, the yield under such conditions is not high. To induce heavy cropping, the trees should be given proper cultural treatment as in the case of other fruit trees.

Irrigation, if available, should be applied in the dry season. The quantity of organic manure and fertilizers should be applied on the basis of age and growth of the tree. The trees should be trained with a clear stem of about one metre and branches allowed to develop above this height. Pruning consists mainly in removing the crossing and superfluous branches.

Storage

Chestnut fruits are spoiled during improper storage. The storage problem has been partly solved through the selection of new cultivars having nuts resistance to fungal spoilage. The new cultivars-crane and orrin have been recommended because of excellent keeping qualities of the nuts. However, the chestnut industry is still based mostly upon the planting of seedling trees. Until growers plant improved cultivars with nuts resistant to spoilage, the storage problem will remain.

It was formerly considered necessary to cure the

nuts before storage. Tests have shown that nuts may be placed in storage immediately after harvest as long as the containers are left partially opened to allow condensation to escape. Curing or drying of the nuts is important to prepare them for eating. The nuts should remain in a dry room temperature for at least four days or until they become partially soft. Some of the starches are then changed to sugar which makes it more tasty to eat.

Varieties

High-yielding strains with ability to produce an attractive colour, size and shape of nuts belonging to four major groups, viz. *Castanea dentata* (American), *C. mollissima* (Chinese), *C. sativa* (European) and *C. crenata* (Japanese) are being cultivated commercially throughout the world. The principal Chinese cultivars are : "Abundance", "Hemming", "Crane", "Culing", "Nanking", "Orrin". These have been adopted widely due to their high quality nuts. The they require short chilling as compared to European chestnuts. Many hybrids have also been evolved as a natural cross between Chinese and American chestnuts.

In India, no named varieties are under cultivation. New planting invariably consists of seedling trees. In Himachal Pradesh two varieties – Marron de Lizan and Spanish Lady – are under cultivation. In Pithoragarh district of U.P. hills, superior strains of Japanese chestnut are under cultivation particularly in the Berinag area.

Pollination

The chestnut is a monoecious tree bearing female and male flowers separately on the same tree. The functioning depends upon the proper fertilization of female flowers. Solitary trees sometimes produce no fruits which shows that they require cross-pollination. It is, therefore, necessary that at least three to four seedling trees are planted near each other, and if trees are budded or grafted, they should be of three or four varieties. Even in such cases, a fairly large number of fruits are found blanks or containing underdeveloped nuts. This may be due to improper pollination and fertilization. In the USA, Dulaney, Griffin, Hathaway and Ketcham are the varieties generally reported to be grown together.

Harvesting, Yield and Marketing

In India, the nuts ripen in October-November. At this time, the bur splits open exposing the nuts inside. The bur along with nuts fall down and are collected from the ground; those persisting on trees are knocked down by poles. The bur still enclosing the nuts are removed by pressing them manually. The yield per tree varies from 10 to 20 kg. The trees begin to bear at the age of 10 to 12 years and go on bearing good crops for 70 to 80 years.

The nuts should be marketed soon after harvesting as with long storage, their quality deteriorates as also mentioned earlier. The nuts are roasted on a slow fire before eating. The kernel is a source of milk food.

Uses of Chestnut Over the World

In many European countries, chestnut is looked upon as a staple diet. In France, from the Bay of Biscay to Switzerland, there are large plantations and almost wildings of chestnut trees. The nuts are large and broad and are extensively eaten by human beings and animals. The poor people during antumn and winter often make two meals daily from chestnut. The ordinary way of cooking them is to remove the outside shell, bleach them and put on fire to steam; they are often eaten with salt or milk, mostly as breakfast by working people.

The nuts are also eaten as a vegetable and are very popular on the table of the well-to-do and wealthy. They are served not only boiled but roasted, steamed, fumed and as dressings for poultry and meat.

Chestnuts are made into bread by the mountain peasantry. After the nuts have been bleached, they are dried and ground. From the flour, a sweet, heavy, flat cake is made. It resembles the Oaten cake so popular amongst the peasants.

In Italy, the chestnut forms a considerable part of the diet of the people during autumn and winter, where they are generally eaten roasted. They are also much eaten in the cooked stage, often prepared like a stew with gravy.

In Korea, by far the most common food nut is the chestnut. It is used as raw, boiled, roasted, cooked with meat, made into confectionary, hardened and mixed with candy and dried whole in which state it becomes quite sweet but is apt to be affected by worms.

In Japan, great quantities of chestnut are raised, but they are used less for human food than elsewhere. They are fed largely to swine. The latter practice may account for the poorer quality of the Japanese chestnut, quantity rather than quality being the consideration among the Japanese. However, large quantities are sold for human food in Tokyo.

Chestnut has been used in food combination in

Asia, Europe and America since recorded history. Formulas have accumulated in history books and cook-books in many languages so that it is impossible to give usual credit to those who first used them. Over 30 old and new methods of using chestnuts are known which provide a choice for serving them in the home, in public eating places and by manufacturers.

7

HAZELNUT – FILBERT (Corylus avellana Linn)

HAZELNUT filbert (Corylus avellana Linn.) are apparently used interchangeably to include all plants in the genus *Corylus*. Hazelnut dates back to ancient times. Old Chinese manuscripts indicate its use as also ancient agricultural history going back nearly 5,000 years (Peker 1962). Filbert culture was known to ancient Greeks and Romans and the nut was supposed to have medicinal qualities and as much used as a food item.

Italy, Spain, Turkey and the USA produce most of the hazelnuts with Turkey giving three-fourths of the world tonnage.

There are two main groups of this nut: filbert which is long and somewhat similar in shape to the normal human finger nail; and cobnut which is roundish and sometimes flattened. The former is in fact an improved version of the latter.

Wild forms of hazelnut are widely distributed in the northern temperate regions of India. The tree is found growing in forests from Kashmir to Kumaon (U.P.) at an altitude of 1,600 to 3,000 m above mean sea level. The nut is called "findak" or "bindak" in Hindi. In Punjab, it is called *Bhangi* or "Urni" and "Bhotia badam" in the Kumaon hills.

The hazel trees growing in nature being a seedling origin show great variation in growth and fruit characters. The tree may be a bush with a height of about 3.6 to 4.5m or may grow to a height of as much as 12 metres. The nuts grown in Shimla hills are smaller in size than the European varieties but are nearly as good in quality. The nut from Kashmir almost resembles the nut from Europe.

Climatic Requirement

Hazelnut thrives in a cool, temperate climate with a moderate rainfall and very little late frost. The late frost kills all the pistillate and staminate flowers. Only the Himalayan regions are suitable for its cultivation as the tree cannot be grown successfully in the plains. Partially shaded gullies found in the hilly areas are especially suited for this nut.

Soil Requirements

Hazelnut plants thrives well in deep, fertile and well drained soils. Waterlogged soils pose adverse effect on the root and shoot development ratio and often cause root and collar rot problems. The soil must be free from hard pan.

Propagation

Hazelnut trees are raised from seed and exhibit great variation and are, therefore, not suitable for use in commercial plantations. The seedlings are mostly used as rootstocks for budding and grafting superior varieties on them.

Varieties

There are no standard named varieties under cultivation in India. The trees are mostly of seedling origin. The important varieties in the USA are

Barcelona and Duchilley. Other good varieties are Cosford Cob, Pearson's Prolific, Atlas, Hempless and Tomollo.

Planting

Hazelnut trees are planted either in regular plantations or as wind-breaks and hedges. When planted as a hedge, the distance from tree to tree is maintained at 180 to 240 cm. In the orchard farm, the trees are planted by the square system with a planting distance of 4.5 metres accommodating about 480 trees to a hectare. The best time for planting is late autumn.

After-care and Cultural Operations

Hazelnut trees should be trained on a single stem up to a height of about 60 cm from ground level and to train the tree on a single stem all branches which emerge below this height should be removed. The suckers coming from the basal portion of the plant should also be cut with a sharp knife as near the point of emergence as possible, unless they are required for producing plants for further planting.

Diseases and Pests

The tree is fairly resistant to disease and insect pests and does not suffer from either of them seriously. Nut weevils are reported to be serious in other countries.

Irrigation

As this nut is grown at high elevations and in fairly moist locations, artificial irrigation is not generally necessary. In warm regions and light soils, irrigation if available should be applied. It will prove useful in proper filling up of the nuts and increasing their size.

Pruning

In the early years, pruning aimed at developing an open-centred tree of medium height with abundant lateral growth is fairly hard. With the bearing trees, pruning is delayed until the male flowers have completed the shedding of their pollen in the spring; otherwise, many catkins are removed before the female flowers are pollinated. The nuts are borne on two-year-old or older spurs on the lateral branches. As the tree grows older, the surplus and older laterals should be thinned out and finer ones retained. All suckers must be removed. The main object of pruning is to cut away sufficient old wood to ensure an ample new twigging fruiting wood.

Pollination and Fruit-setting

The hazelnut is a monoecious plant. The male flowers are in a conspicuous cluster or catkins. They bear profuse pollen which is dispersed by the wind in early spring. The female flowers are not so prominent. They appear only as prominent buds from which the red stigma protrudes about the same time the male flowers emerge. Blossoming begins in March or early April and lasts for four to six weeks. Low yields may sometimes be due to damage to catkins, high wind blowing away pollen or the female flowers becoming receptive after the pollen has been shed.

Most varieties are self-incompitable and require cross-pollination. However, planting of two or more varieties such as Cosford, Duchilley, Barcelona, Notingham, Du Province, Brixnut, Nook Sack, White Aveline and Doviana is generally satisfactory from this point of view. No evidence of cross incompatibility has been reported in this nut.

Harvesting

It may be commenced as soon as the husk enclosing the nut becomes brown. However, the general practice is to defer harvesting until the nuts start falling from the tree. At this time, the husk can be readily removed. Harvesting is done by hand-picking, or the trees may be shaken or the branches struck with pole to cause the nuts to fall on the ground. A canvas sheet may be placed under the tree to collect the nuts. They should then be picked up and dried for a short time in the sun or in an airy shed. Remaining husk should be removed and nuts collected. The bright colour of the nuts can be retained by mixing them with common salt. In certain seasons, a certain proportion of nuts may develop without any kernels, which is due to frost injury or defective pollination. The nuts can be stored for a long time in tight jars in a cool and dry place.

Peeling

Roasting the filberts facilitates their peeling. The optimum on roast has been obtained by roasting the nuts for three-four minutes at 148° to 155°C in coconut oil. The best oven roast is obtained at 205°C for seven minutes.

Yield

Hazelnut trees do not bear any fruit for the first five or six years. After attaining an age of six years, a crop of about 2.25 kg per tree may be produced. The average yield from a full-bearing orchard is from 625 to 1,125 kg per hectare. The highest yield is obtained at the age of 15 years. The nuts are ready for picking in August-September. The trees are fairly reliable in cropping and are long-living. Many trees in Europe are regarded to be bearing profitable crops even after 70-80 years.

World Production

Turkey leads in production of hazelnut accounting for 70 per cent world production in 1984-88, followed by Italy which accounts for 24 per cent, with Spain and the USA occupying the third and fourth positions. (See Table 7.1).

TABLE 7.1: World hazelnut crop

(M. Tonnes in Shell)

Country	1984	1985	1986	1987	1988
Italy	55000	130000	90000	90000	150000
Spain	12000	32000	27000	30000	30000
Turkey	300000	230000	320000	300000	430000
USA	11800	22300	12000	20000	15500

World Import

West Germany is the largest importer of hazelnut followed by France and the USSR.

Table 7.2 : Gives the imports of the major consuming countries

TABLE 7.2 : Imports of hazelnut

(M. Tonnes)

Country	1984	1985	1986	1987	1988
Western Europe					
Austria	7535	7067	7855	7803	8865
Belgium	5655	6413	5614	6336	6500
Denmark	2119	1955	1992	1970	2302
Finland	226	303	259	159	236
France	17175	16938	15646	16131	15781
Germany (Federal Republic)					
(in-shell)	3654	5232	3657	4209	334
(kernel)	62674	61129	57549	61751	58798
Netherlands	6677	5783	5000	5937	6659
Norway	2363	1892	2158	2176	1961
Sweden	4797	4664	4357	4273	4139
Switzerland	12185	10825	12370	11445	12944
U.K.					
(in-shell)	1379	1462	1321	2054	-
(kernel)	9228	7919	7836	5771	-
USSR	34580	12600	7742	15478	-
America					
Canada					
(in-shell)	839	662	604	857	1303
(kernel)	1082	1044	1034	838	1119
USA	4171	3579	1467	1264	1252
Asia and Oceania					
Australia	1549	1485	1469	1265	1252
Japan	582	462	561	648	328
New Zealand	71	68	79	74	96

Figures not available.

Note: Unless otherwise stated figures include in-shell and kernels.

Source: Menninger, E.A. (1977), Edible Nuts of the world, Horticultural Books, inc. P.O. Box 107 Stuart, Florida, USA.

World Exports

Turkey leads in exports as in production followed by Italy (see Table 7.3).

TABLE 7.3 : Export of hazelnut

(M. Tonnes)

Country	1984	1985	1986	1987	1988
Italy					
(in-shell)	9248	10482	6102	6607	7000
(kernel)	23944	29553	17478	20315	30000
Spain					
(in-shell)	69	155	771	1443	374
(kernel)	5713	4378	4050	6289	5608
Turkey					
(in-shell)	5832	107225	136858	132214	157197
(kernel)	135542				
USA					
(in-shell)	1322	1463	3718	4012	2220
(kernel)	663	1187	3241	1592	1394

8

PECAN NUT
(Carya illinoensis Koch.)

PECAN is the most favourite edible nut. It belongs to the family of Juglandaceae. Out of 20 species of genus *Carya*, *Carya illinoensis* is extensively cultivated on commercial scale in the USA, Mexico, Australia, Canada and Western Europe. The other members of this family are Pig nut (*Carya blabra*), the small pignut (*C. ovalis*), the Shell Bark (*C. laciniosa*) and Shag Bark Hickories (*C. ovata*). Except for the pecan, the other close relatives are grouped under Hickory.

Pecans are native to north America. The natural range extends from the north-central USA to northern Mexico, though due to distinct genetic and physiological adaptations, the genus is found in a vast geographical range with different temperatures and humidity levels.

The cultivation of the pecan nut in India is still in the initial stages. The limited number of trees that are fruiting in this country are exotic. Seedling trees do not yield true-to-type nuts. Attention is, therefore, required to the raising of high quality nursery stock through vegetative propagation.

Climatic Requirement

The pecan tree thrives best in locations which are free from severe frost in spring and excessive heat during summers. In India, it can be grown successfully in the low and mid-hill elevations of about 1,200 to 1,600 m above sea mean level in the Himalayan region. Areas with heavy rainfall are not so suitable as those with a moderate rainfall of about 75 to 100 cm. Its cultivation can be extended to zones which are considered somewhat lower and warmer for the walnut.

Soil Requirement

The pecan tree is deep-rooted and needs fairly large quantities of water throughout the growing season. Soils adopted for pecan production should be fertile, well-drained, deep and capable of retaining adequate moisture. Alluvial soils, if deep and well-drained, are usually satisfactory. Upland soils are usually too shallow and droughty for commercial pecan production. The pH range for pecans should be in between 5.6 and 6.5. If the soils are acidic and show low pH the soil should be reclaimed through dolomitic limestone application. The planting on badly eroded hills, bottom lands or pockets should be avoided as these sites do not permit free air movement resulting in slow drying due to excessive humidity.

Propagation

Pecan seeds lose their viability too early if the nuts are stored under dry and warm conditions. To prevent this loss they should be stored at 0°C immediately after harvest. For obtaining good and rapid germination the seeds are stratified at low temperatures (1 to 5°C) for about 12 to 16 weeks (Staden *et al*, 1976; Dimalla and Staden, 1978).

The seeds which are not subjected to

stratification treatments exhibit poor seedling growth and require high temperatures of 30 to 35°C for germination. In order to ensure good germination and to facilitate stratification treatments at low temperatures the seeds should be sown under natural field conditions in December-January. The seeds can be sown in polythene bags also, because the seedlings grown in polythene bags can easily be planted in the desired site with earthen balls without causing any damage to their roots which is very often the case with the seedlings raised in the nursery, only deep, well-drained sandy soil should be used. The seeds should be soaked in water for about 10 hours prior to their planting. In order to avoid damage caused by rodents and birds, the seeds should be painted with a paste made of red lead oxide and linseed oil. Pecans can also be propagated vegetatively through budding or grafting selected scion cultivars on pecan seedling root-stocks (Bryant, 1969). "Patch-budding" can be done in the month of June while "Tongue-grafting" is done during February. Excellent rooting in 50-cm-long and 20-mm-thick hardwood cuttings treated with 10,000 ppm of indole butyric acid (a root-promoting hormone) is also reported in pecan nut (Smith *et al*, 1974). Full-grown trees producing inferior nuts can be changed to the desired cultivar by crown-grafting in late winter or early spring. The seedlings can be out-planted at the age of two – three years. For ensuring better survival at the orchard site the seedling should attain a height of at least 120 to 130 cm.

Varieties

Variety selection may be a key to successful planting. Disease resistance is a major consideration in areas experiencing high humidity and poor air circulation. Besides this, characteristics like tree size, productiveness, shell and kernel quality and nut appearance should also be considered while selecting a pecan cultivar for commercial orchards. Some of the promising pecan cultivars are listed below :

(i) *Desirable*

Produces large nut with a semi-thin shell which cracks easily. Attractive kernel, being a heavy pollen-shedding variety. The planting of this variety in the orchards with the other varieties facilitates pollination. The tree is vigorous in growth. Regular bearer.

(ii) *Nellis*

Nuts are long and shells are thin. Kernels are of good quality.

(iii) *Mahan*

Produces very large nuts with a fairly thin shell. Good variety for the areas experiencing abnormal rainfall.

(iv) *Burkett*

Large round nuts with a semi-thin shell which is not easily separated from the kernel.

(v) *Stuart*

Produces large attractive nuts with well-defined markings and a fairly thick shell. Shell easily crackable. Moderate bearer.

(vi) *Elliott*

Small round nut with medium thick shell and good flavour. Scab resistant. Early maturing. The trees are vigorous but slow to attain large size. Consistent bearer.

(vii) *Curtis*

Small thin-shelled nut suitable for shelling, high quality kernel producing ability. Moderately resistant to scab. The variety is suitable for lower elevations experiencing a long growing season.

(viii) *Cape Fear*

Fairly large nuts. Shell cracks well and possesses high quality. Trees are highly productive and start bearing fruit after seven to eight years of age.

Pollination

The male and female flowers respectively known as catkins and pistillate flowers are borne on the same tree separately. The pollen-shedding is facilitated by winds. Blooming periods differ from variety to variety, consequently fertilization does not take place properly either for want of pollen or due to unreceptiveness of stigma. This results in poor fruit-setting and sometimes causes premature losses of nuts due to inadequate fertilization. Pollen is produced in great abundance by the catkins and is distributed by gravity and wind. The pollination is adversely affected by high humid conditions, therefore, sufficient daylight with a peak in the afternoon should be available during this process.

The fertilization process takes about five-seven weeks after pollination. Maturity of nuts begins with the development of shell size and kernel. During this time the weight and specific gravity of the fruit increase rapidly. Water content decreases from about 75 per cent to 30 per cent. The amount of fat becomes seven times greater as compared to the initial stages of fruit development. The hull turns from pale green to greenish black and cracks at full maturity.

Planting Season and Technique

Young nursery stocks can be transplanted any time during the dormant season but planting from late January to the first fortnight of February has been found most successful. The pits of 1×1×1 m size should be dug at a distance of 10×10 m. Each pit is filled with a mixture of compost, soil and 150 gm of Aldrin 5 per cent dust. The filled soil surface should be at least 10 cm above the normal soil surface. Before setting, roots of the trees should be examined and all broken or injured roots removed or cut back to eliminate the injured portion. To prevent drying of roots, the roots are dampened in burlap while the plants are carried to the actual planting site. If possible, transplanting should be practised on a cloudy day preferably in windless and high humidity conditions. In order to achieve high survival and proper root development the soil must be adequately moist to facilitate root development. Mulching around the tree basin helps in conserving the soil moisture for a longer period. In order to control the injuries caused by low temperatures or sun scaled or borer damage to the newly planted plants, the main trunk should be painted up to 60 cm height from ground level with white zinc paint.

Training and Pruning

At the time of planting one-third or one-fourth of the top portion should be cut so as to balance the shoot and root ratio. On the main trunk the lowest branch should be kept at a height of 1 m from the ground level. The space between the main limbs should be maintained at 35 to 40 cm. Pecan trees do not require more pruning at later stages. However, during the initial stages (up to five-six years) light pruning is essential in order to remove the weak, diseased, interlocked branches. The lateral limbs should be allowed to grow equally around the main trunk of the tree. Once the framework is established very little pruning is done for removing the weak, splitted, weeping, diseased or dead branches.

Manuring

To realize good annual production, trees must be judiciously nourished. Pecan trees need a supply of all the major elements as well as few of the minor elements like zinc, manganese and boron. Nitrogen, phosphorus and potassium are the nutrients needed most. Out of these three, nitrogen is the most important to regulate the proper growth and development of the trees. On an average, annual terminal twig growth of 25 to 30 cm is required in order to maintain the productivity of the trees for a longer period. A fertilization programme for bearing and non-bearing pecan trees will depend largely on soil management practices and soil fertility. In order to improve the soil, well-decomposed 50-60 kg of farm yard manure or compost should be applied every year during December-January. On the basis of per year age of tree, apply 0.5 kg of NPK mixture containing equal parts of nitrogen, phosphorus and potassium in 15:15:15 ratio. The quantity of the fertilizer will increase accordingly in the arithmatic progression every year up to the age of 16 years. The dose of fertilizer fixed for adult trees (8 kg of NPK mixture 15:15:15) will be applied to the trees which are older than 16 years. The proper time to apply fertilizer is early spring. Pecan trees suffering from zinc deficiency exhibit drying of apical shoots, little leaf forming in a clustre (known as rosette) bronzing and mottling of leaves, early defoliation, ultimately resulting in poor longevity of the trees and also affecting nut yields adversely. If any symptom of zinc deficiency is noticed on the trees apply foliar spray of zinc sulphate at the rate of 1-2 kg per 500 litres of water. To avoid toxic effects of sulphate ions (SO_4) on the tender plant tissues, the activeness of the chemical must be neutralized by adding slaked lime at the rate of 0.5 kg per kilogram of zinc sulphate. Sometimes pecan trees also exhibit blossom splitting, poor filling of the nuts and gummosis – these physiological disorders often take place due to boron deficiency. Growers are advised to apply 0.5 kg of borax (commonly known as Suhaga) in the soil of tree basin.

Early Care

Watering during dry time is essential for the survival of the pecan tree throughout the first year. Only a time period of drought and crowding by grass or weeds can kill the trees in the first summer. Therefore, water the trees when rainless periods exceed 7-10 days. Drip irrigation can be used effectively for establishing an orchard. This increases the survival rate and allows more growth in the first years. Over watering of a pecan tree should be avoided as excessive moisture damages it.

Alternate Bearing

Alternate pecan production (on and off years) is mainly the result of inadequate fertilization. When trees set a large nut crop, and nutrients and moisture are not enough for the nuts to mature and for the tree to store enough plant food, production will be low the following year. Early defoliation in the autumn usually means no nut crop the next year. Diseases and insects affecting the leaves also contribute to alternate bearing by causing early leaf drop in the autumn.

To help prevent alternate bearing, use sound cultural practices. These include disease and insect control, adequate use of fertilizer and zinc sulphate, and an extra application of fertilizer in late May or June in years when nutset is heavy. And keeping healthy, dark green leaves on the tree is a sure step toward achieving annual nut production.

Factors Influencing Yield

The most important factor influencing yield is proper fertilization, including zinc, to prevent rosette. Controlling diseases and insects is equally essential, along with proper spacing of trees for root growth and adequate sunlight.

Unfavourable factors are late-spring frosts or freezes, droughts, and storms as they injure the trees and reduce production. Storms and sleet also cause limb-breaking that reduces crops for a few years. Strong winds may also cause loss of the current crop of nuts.

Why Nuts Fail to Fill

Sometimes nuts may fail to fill. This is caused mainly by insect and disease damage to leaves and an inadequate number of leaves. Early defoliation caused by disease, insects or deficient nutrition directly influences filling.

Drought, if it occurs in the latter part of the growing season, also causes failure to fill because the transportation of nutrients is disrupted. The shell is formed earlier and filling occurs later in the season.

Premature Loss of Nuts

Lack of pollination causes the great loss of nuts. Since pecans are wind-pollinated, excessive rain during bloom prevents pollination and the unpollinated nuts fall. Weather conditions in some seasons cause the male and female flowers to mature at different periods, and pollination fails to occur. Some male varieties shed their pollen before the female flower is ready to receive.

Insect damage also can cause premature loss of nuts. Loss is also attributed to black aphids, pecan weevils, stink bugs, and shuckworm.

Scab, mildew, blotch, and other diseases also cause nut loss. Scab is usually bad if rainy, humid conditions occur during the early part of the season.

Inadequate fertilization and drought often cause early drop of nuts.To prevent this, reduce competition between weeds and grass.

When trees have been neglected or planted in unsuitable soil, loss of nuts is often rather severe.

Harvesting

Pecans are harvested as soon as nuts separate from the shucks. In small plantings, the kernels can be thrashed with cane poles. The pecans drop on spread canvas or bare ground where they are picked up by hand. Pecan nuts lose quality very quickly on the ground, especially during wet weather. Poorly filled pecans are especially subject to rapid loss of quality. Occasionally, nuts ripen and the shucks fail to open sufficiently for doughing of the kernel. Sprouting occurs especially during wet weather. The "shuck-held" pecans are shaken down from the tree. Shucks then open on the ground.

Yield

Compared with other tree fruits, pecan trees are rather slow to come into production. Even on varieties known to produce early, such as Desirable, Curtis and Elliott, only a few nuts will set fruits in two to three years. Stuart usually requires four or more years. And in most cases trees are 10 to 12 years old before production becomes commercially importancc. Well-grown trees may reach this production at eight years of age. However, from about 10-year-olds to about 20-year-olds, one can expect production to be about 0.5 kg per year of age of the tree. After this age, production per tree increases rapidly. Well-cared-for orchards of mature trees should produce around 2,500 kg per hectare per year, but the average annual production is usually not so large. Some large trees in good condition have been known to produce as much as 150 kg per tree.

Curing

If pecans are harvested during wet periods, the nuts should be sacked and stacked openly or spread shallowly in well-ventilated areas until dry. Otherwise, sprouting or moulding might occur in storage. If harvesting is done in clear, dry weather, the pecans can be sacked in cloth bags and placed in ventilated areas for curing.

Storing

When the crop is stored, the nuts should be placed in a cool, dry, protected place. Temperatures of 1° to 3°C are ideal. Higher temperatures are satisfactory for short storing periods. For storing for longer than one year freezing or below freezing temperatures are necessary.

Market

The improved "paper shell" varieties are sold in the USA for shelling or packaging as whole nuts, while the native pecan are generally sold for shelling purposes. According to a survey by the U.S. Department of Agriculture of the pecan industry, less than 10 per cent of the national pecan crop is sold and used "in shell". The rest are shelled. Forty-two per cent of the shelled crop goes to bakeries and ice-cream manufacturers, 19 per cent to confectioners and 33 per cent to retailers.

9

PISTACHIO NUT
(Pistacia vera L.)

PISTACHIO belongs to the family Anacardiaceae to which other fruit specie, viz. mango, cashew nut, sumac species in the genus found in Asia Minor, eastern Asia and southern North America also belong but *P. vera* has been selected for commercial purpose due to its high kernel value. The remainder species are either used for rootstocks (*P. terebinthus* and *P. atlantica*) or for ornamental purposes (*P. chinensis*). Pistachio is supposed to have come from Syria and a region eastward as far as Turkestan. It has been known for perhaps 3,500 years in western Asia and for more than 1,500 years in Italy. It is grown, on commercial scale in Iran, Turkey, Afghanistan, the USA, the USSR, Syria, Greece and Italy but cultivated rather extensively in Sicily. In America it is largely used for flavouring icecream. Pistachio tree has the resinous odour which it inherited as a distinct trait of the family Anacardiaceae. The nut is smaller in size than other temperate nuts yet it is largely demanded due to its rich flavour and nutritive value. It bears 53.7 per cent oil, containing fatty acids like myristic acid 0.6 per cent; Palmitic acid 8.2 per cent; Stearic acid 1.6 per cent; Oleic acid 69.0 per cent; Linoleic acid 19.8 per cent; and unsaponifiable matter 0.8 per cent. The oil has a dark green colour and an aggreable odour like the nut (Jacobs, 1951). However, the composition of the pistachio varies from cultivar to cultivar and parts used as is apparent from Tables – 9.1 and 9.2.

TABLE 9.1: Composition of pistachio nut kernels, shells and hulls (per cent dry basis)

Composition	Kernels		Shells		Hulls	
	Split nuts	Unsplit nuts	Split nuts	Unsplit nuts	Split nuts	Unsplit nuts
Protein NX6.25	19.41	19.58	0.42	1.06	7.66	9.35
Oil, ether extractable	58.30	54.70	0.56	0.58	7.82	8.27
Crude fibre	1.74	2.19	54.00	53.40	14.10	17.40
Ash	2.95	3.55	0.42	1.06	15.56	13.30
Calcium	0.13	0.13	0.06	0.06	0.07	0.25
Potassium	1.04	1.22	0.22	0.49	6.71	5.88
Phosphorus	0.54	0.64	0.02	0.04	0.12	0.24

Source: Bloch and Brekke (1960)

TABLE 9.2: Composition of pistachio nutmeant from split nuts (per cent dry basis) of different cultivars*

Composition	Cultivar			
	Red Aleppo	Kerman	Trabonella	Bronte
Protein (NX6.25)	19.00–23.00	26.00–28.00	24.13–24.25	24.81
Oil, ether extractable	54.00–58.00	48.00–50.00	54.00–55.00	53.20
Crude fibre	1.65–1.74	1.93–2.11	1.74–1.97	2.11
Ash	2.95–3.34	2.95–3.50	2.85–2.90	3.03
Phosphorus	0.54–0.57	0.54–0.64	0.49–0.50	0.55
Calcium	0.13–0.16	0.10–0.11	0.11–0.18	0.18
Magnesium	0.17	0.15	0.17	0.18
Potassium	1.04–1.17	0.93–1.18	0.86–1.06	0.92
Total sugar (as sucrose)	7.2–7.5	7.9–8.4	6.1–6.2	6.4

* Range based on three-year data.
Source : Bloch and Brekke (1960).

Tree Morphology

Pistachio is a deciduous tree of dioecious nature (producing staminate and pistillate flowers on the different trees). The adult trees attain a height of about five metres like peaches and is a slow grower. The pollination is completed by wind. The maturity of anthers and receptivity of stigma varies from cultivar to cultivar, thus usually exhibits alternate bearing and sometimes unfruitfulness. Botanically, the fruit is classified as *oblique ovoid drupe*. Nuts and kernels have resinous odour which is the inherited character of genus Pistachio. Kernels are edible.

Climatic Requirements

P. vera trees can tolerate quite low temperatures and require longer chilling requirements. Warmer winters often delay the opening of the buds. Pistachio is thought to survive drought and other adverse soil and climatic conditions better than walnut and almond. It thrives well in hot valleys and desert climate where winters are cold enough to fulfil the chilling requirements for proper breaking of bud dormancy. Similarly long hot and dry summers are needed for the proper development and maturity of the nuts.

Excessive humidity results in a high degree of incidence of disease and insects and pests, thereby effecting nut yields adversely. Although pistachio trees appear to be drought tolerant they cannot withstand prolonged droughts during summers and require adequate soil moisture for proper plant and nut development. Under severe moisture stress conditions the leaves of the trees start discolouring from green to yellow and drop before crop maturity which in turn influences the nut yield adversely. During extremely hot summers, if possible, two to three irrigations at an interval of 20 days should be given to avoid premature leaf fall and leaf discolouration and promote sound filling of nuts. The root systems of pistachio are easily affected by root rot/wilt if soil wetness persists for longer periods. Therefore, the sites subject to high water table or waterlogging should not be selected for its cultivation.

Soil Requirement

Pistachio has the ability to grow in a wide range of soil conditions. Dry hilly sites with rocky soils have proved healthy for its survival and growth as is apparent from natural stands growing on gypsiferrous rocky soils of Lebanon, Palestine and the Isle of Cyprus. Loamy or sandy loam soils rich in gypsum and carbonates or crystalline calcareous, well-drained and slopy sites are also suitable for its cultivation. For obtaining higher survival, growth and nut yields pistachio nut should always be cultivated on deep, light, well-drained and fertile soils having a minimum depth of 80-95 cm.

Propagation

For achieving higher survival from out plantings, pistachio seedlings should always be raised in polythene containers, the reason being the poor adaptability of roots to planting site, if used as bare root seedlings.

The seeds should be collected on turning of the hull colour from blue to green. The hulls are removed from the kernels otherwise these will hinder germination as they contain inhibitors. Prior to their sowing in containers pistachio seeds require stratification and, therefore, they should be placed in moist sand for about four-five weeks at a low temperature (5°C). One seed should be sown in each container. Seeds sown in January-February under indoor conditions are placed outside in March so that the young seedlings may grow faster under full sunlight. Pistachio seedling attains sufficient height and stem thickness in 15-18 months and can be out-planted at their permanent growing sites without damaging the roots and earthen ball. Pistachio nut can also be propagated vegetatively by budding on seedling rootstocks. Seedlings of *Pistacea terebinthus* and *P. atlantica* are extensively used as rootstock in California. Seedling of both the species are resistant to nematodes (Joley and Whitehouse, 1953). For propagating pistachio through budding "T" method is adopted in midspring, when sap flow is excessive. In the areas more prone to verticillium wilt, seedlings of *P. integerrima* should be used as rootstock as this species bears a fair degree of resistance to verticillium wilt.

Commercial Cultivars

Red Aleppo, Trabonella, Kerman, Bronte are recommended for high quality pistachio nuts and these cultivars should be planted with Peters or USDA 23 which are good pollinizers.

Planting

The seedlings or budded plants may be planted at a distance of 6×6 m (plant × row). Planting distance on fertile soils can be increased slightly from 6×6 m to 8×8 m to avoid any crowding. Being a dioecious species, male trees are spaced throughout the orchard for ensuring good pollination and setting of fruits. One male (staminate) tree should be planted for every six female (pistillate) trees to ensure adequate pollination and nut yield. Transfer of pollen to pistillate flowers is completed by winds.

Training and Pruning

To provide an ideal framework, the trees at their initial stages of growth are trained in modified leader system. For training of pistachio trees with this system the main stem is allowed to grow with evenly distributed three – five primary scaffold branches at a spacing of 30 cm from each other. The height of the lowermost branch is kept at 90 cm from ground level. The scaffold limbs are trained at an angle of 45 degrees to avoid splitting of limber during strong winds. The centre of the tree is kept open to expose the maximum tree area to sunlight which aids adequate flower bud, formation and subsequent fruiting. After training the trees in a desired shape, very light pruning is done in the adult bearing trees, like cutting of a limited number of lateral vegetative shoots which is essential to maintain the desired vegetative and reproductive balance so as to make them productive for a longer period.

If pruning is not done regularly, the bearing area extends farther and farther away from the centre of the tree; as pistachio produces its flower buds laterally on the shoots. Thereore, new lateral growth is essentially stimulated each year by doing regular light pruning. Heavy pruning shall be avoided as these heal very slowly.

Manurial Requirements

Mature bearing trees should be fed with 50 kg of well decomposed compost and three kg of NPK mixture (15:15:15). In order to fulfil the needs of trace elements foliar spray of multiplex at the rate of 0.3 per cent concentration should be done during April-May.

Diseases, Insects and Pests

Verticillium with (*Verticillium alboatrum*) is a serious disease of pistachia. In the problematic sites which are prone to Verticillium, *Pistachio integerrima* seedlings which are resistant to this disease should be used as rootstock. Before planting the soil should be sterilized. Crown rot (*Phytophthora parasitica*) is another disease which causes high mortality to pistachio. *P. terebinthus* is supposed to be immune to the fungus and should be used as rootstock. The younger seedlings very often suffer injuries due to infestation of defoliators, root and bark-eating caterpillars and, therefore, timely action with suitable insecticide should be taken.

Harvesting and Yield

Having the biennial bearing tendency pistachio trees bear heavy crops in one year followed by a light crop in the next. The nuts can be harvested manually as well as mechanically. For mechanical harvesting the trees are kept low and vase-shaped. Nuts are harvested on the loosening of shell. After harvesting the nuts are placed in bags. Outer husks of a small percentage of the overmature or immature nuts adhere to the shell but can be easily removed if they are water-soaked for 30 minutes. Nuts having undeveloped nuts floating in the water surface should be removed. A full-grown mature tree yields about 15-20 kg of marketable nuts annually.

Drying of Nuts

With a view to preventing spoilage in the harvested nuts, the pistachio is dried immediately after harvest. Pistachio nuts are dried with forced air at 35° to 37.5°C. The moisture content of the freshly harvested nuts is maintained at 45 to 50 per cent by forced air-drying.

Grading

After the sorting out of discoloured or blemished nuts, classification is done into four major grades by squirrel caze sizer. The nuts are stored at low temperatures till they are salted, roasted and packaged.

Salting and Roasting of In-shell Nuts

For the consumption of pistachios in the form of "snack" the nuts are salted and roasted before they reach the consumer. But for inducing flavours, confectionary and ice-cream the nuts are sold unsalted and unroasted. For salting, the nuts are passed rapidly through a saturated brine solution, and then dried in a rotary-screen drier at 71°C for 30 minutes till they are free from the moisture absorbed by the nuts during the salting process. These moisture-free salted pistachios are then exposed to 120°C for 10 minutes.

Uses

Pistachio nut is used in several forms for various products, like ice-cream, confectionary and meat dishes. Ninety per cent of the in-shelled pistachio is used as roasted and salted for snack purposes, and the remaining 10 per cent is used as flavouring confectionary items and ice-cream in the form of nut meats.

10

CHILGOZA PINE NUT
(Pinus gerardiana Wall.)

CHILGOZA pine is a conifer of compact appearance with short lateral branches. It belongs to the family pinaceae, and has its origin in the north-western Himalayan mountains. It is a close relative of other pine species of the Himalayan region. The three needles of a chilgoza pine differentiate it from the other species. The tree at full maturity attains a height of about 16 to 20 m with varying trunk girth ranging between 3 and 3.8m.

Chilgoza pine is very common in Afghanistan, Tibet and northern Baluchistan where the gregarious tree can easily be seen growing naturally. In India, it covers an area of about 2,022 ha. (Singh *et al*, 1973) and grows naturally in the inner Himalayan zone comprising Kinnaur and Pangi in Himachal Pradesh and the Kishtwar and Astur areas of Jammu and Kashmir. The naturally growing trees in the adjacent forest of the Niti Pass, in Uttar Pradesh, have also been reported.

Climatic and Soil Requirement

Being a native of dry regions outside the influence of monsoon, with very scanty rainfall and heavy snowfall during winters, its cultivation could not be popularized for want of the peculiar climatic requirements. During summers it requires a temperature of less than 35°C for its proper growth. The soils for chilgoza pine should have good drainage and be of light texture allowing adequate aeration for the development of the root zone. Soils with a moderate fertility status are deemed fit for its growth and development. However, the tree has very little demand of the nutrients and is capable of growing on excessively dry, barren hill sites, even on base rocks. Inspite of its Himalayan origin the tree is winter tender at low altitudes and does not thrive.

Flowering and Fruiting Behaviour

The male and female cones are borne on the same plant in May-June. The number of male cones on the plant is proportionately high as compared to female cones. The tree starts producing female cones after attaining the age of 25 to 30 years. The young female cones increase in size after pollination, slightly during the first year but rapidly during the second year, and attain maturity by September-October. From pollination to seed set it takes about two years. The length of the mature cone varies from 12.7 cm to 23 cm with a diameter of about 8 to 12 cm. The seeds are enclosed by very thick woody scales which open to disperse the seed at full maturity. On an average there are 30 seeds in a cone and 4,600 seeds weigh one kg.

Harvesting and Storage

The nuts are ready for harvesting in the month of September-October after a period of 18 months from its blooming. The average yield of about 1.8 kg is obtained from an adult tree. The cones should be harvested before loosening of the outer woody scales which can be easily opened by artificial or solar heat. The nuts are then collected and air-dried for five-six days and stored in a cool place. The nut is cylindrical with a length of 2.03 to 2.54 cm having a short wing which soon becomes detached and remains on the cone when the seed falls.

Nursery Raising Techniques

Chilgoza pine loses its viability too early, if it is not stored at very low temperatures. The optimum time for seed-sowing is October- November, when it can directly be sown in polythene bags. It would be wise to stratify the seeds at low temperatures of 2°C in moist sand for a period of 20 days. The mixture for polythene bags should be prepared by mixing well-decomposed compost, sand and soil in the proportion of 1:2:1. In order to protect the seed from bird and rodent damage, it should be treated with a paint prepared by mixing red lead oxide, commonly known as *sindur*, and linseed oil in 2:1 ratio. As soon as germination starts and cotyledons appear through the soil mixture, they become prey to birds, rodents and small animals. Therefore, the polythene bags should be properly covered by a suitable wire net. Chilgoza seedlings are ready for outplanting after attaining a height of 25 to 30 cm which generally comes after three-four years. Before planting out the seedlings the pits of 60×60×60 cm size should be dug and refilled with a mixture of compost and soil. The planting distance between seedlings should be maintained at 5×5 m.

Uses

The nut is used raw, roasted or in confectionary products. It contains 51.3 per cent fat, 8.7 per cent water, 13.6 per cent protein, 22.5 per cent starch, 0.9 per cent fibre and 3 per cent mineral and ash; it possesses a tremendous wound-healing property.

11

APRICOT
(Prunus armeniaca L.)

APRICOT, a member of the Rosaceae family is supposed to have originated in western China. Few winter hardy strains identified as Russian apricots are considered as native of the Siberian region. The general area of distribution of the wild species of apricot is located in the temperate zone of Asia between 133° and 70°E and 52° and 30°N. The distribution area extends from North Korea and the southern Ussurian region and to parts of Manchuria and north China. Due to its attractive colour, pleasant aroma and sweetness, everyone feels delighted to use it as fresh fruit (epicarp and mesocarp). Depending upon the sweet or bitter taste, apricot kernel is used for oil extraction in many parts of the world and is either used for medicinal purposes or as edible oil, an alternative source. Apricot kernels of some cultivars like Charmagz, Saffeda, Turkey and Large Early Montagamet are pleasant in appearance, flavour and taste like almond kernels and are widely used for consumption as raw, roasted or in the various confectionary products.

Climatic Requirements

Sites located at elevations of 1,400-2,000 m prove suitable for its cultivation. Occurrence of spring frost may cause severe freezing injuries to growing vegetative buds and bloosoms, therefore, the cultivation is restricted to the tracts where there is little danger of spring frost. Frequent fluctuation in the temperature and soil moisture condition also restricts the cultivation of this crop. Higher temperatures and excessive humidity during summers also restrict its cultivation, which makes the condition more conducive for the rapid spread of Brown Rot disease (caused by *Sclerotinia* sp.) resulting in great reduced fruit yields.

The winter hardiness in apricots varies from cultivar to cultivar, based on their genetic constitution and the taste of the kernels. Sweet kernel-producing cultivars with white-skinned flesh require comparatively colder climates than the cultivars producing orange-flesh and bitter kernels. Small sized apricots locally called *Chulu* in Tehi Garhwal Dist. of U.P. Himalayas are very popular mainly for "Chulu Oil" which has medicinal values also.

Soil Requirement

Apricot trees can withstand a wider range of soils but they should be capable of allowing good aeration and to drain out excess water. For this reason, dry temperate zones are ideally suited for cultivation of superior quality fruit-producing cultivars with high sugar content like Charmagz, Saffeda, Sakarpara, Kaisha, etc. The fruits of these varieties are used as dry fruits.

Propagation

In order to raise true to type clonal plant, the scion woods of bearing trees can be grafted (cleft or tongue grafting) on peach and apricot seedlings. The age of the rootstock should not be less than one year with pencil thick stem girth. "T" budding and "ring" budding can also be practised during May-June.

Planting

The grafted/budded or seedling plants with satisfactory vigour are planted during winters (December-January). The pits at a distance of 6x6 m are dug and refilled with a compound of thoroughly mixed compost and soil and 100 gm of Aldrin 5 per cent dust to prevent insect damage. The size of the pit may vary from place to place depending upon the texture and composition of the soil. On the poor degraded soils the size of the pit should be kept as 1 m (1×1×1 m) while on the well fetilized light textured soil, the size of the pit particularly depth may be reduced. The plants should be planted in the centre of the refilled pits and the level of the graft-scion union (in vegetatively propogated plant) should be kept at least 7 to 10 cm above the ground level.

Pollination

In order to maintain the productivity of apricot orchards the plant comprising different varieties with the coinciding blooming period should be planted. Honeybee hives can be placed in the orchard during the blooming period for efficient pollination. Regular pruning and proper fertilization also help in maintaining the productivity of the apricot trees for a longer period.

Harvesting and Yield

Apricot tree starts bearing fruit at five-seven years of

age depending upon the mode of propagation. The fruit can be harvested up to 35 years or more. On an average about 30 to 40 kg of fruit are produced by a single tree. The stone (seed) is extracted from the pulp and washed in fresh-water and is air-dried. The kernels are extracted manually as well as mechanically for their use as oil or raw consumption. The oilcake can be used as livestock feed as well as soil ameliorant.

12

CHEURA
(Aesandra butyracea Roxb.)

BOTANICALLY, cheura is classified as "Berry fruit" and does not belong to the nuts family in the true sense but considering its multiple use (i.e. fuelwood, fodder, feed, edible fruit, oil source, etc.) a serious thinking on its area expansion is required. Cheura is a native of fairly warmer sub-tropical climate and natural scattered stands of cheura are found throughout the outer Himalayan ranges starting from the Kumaon hills and extending to Sikkim, Nepal and Bhutan. In the Andamans also natural stands of cheura are found growing in association with deciduous and semi-evergreen forests. Cheura trees are deciduous in nature and full grown trees may attain a height of 12 to 18 m.

Tree Morphology

Cheura tree has a medium height, and its leaf blade is 20 to 35 cm long and 8 to 15 cm wide. The leaves are in greater numbers near the end of the branches.

Cheura flowers are white in colour with a pleasant fragrance. The diameter of the flower is 2 to 2.5 cm. Blooming takes place in January-February. Fruits are oval, fleshy, edible and mature by June-July, and are grouped botanically under "Berry'. Generally, the legnth of the fruit varies from

2 to 4.5 cm and to contains one or two seeds. Its colour is yellow, has a soft sugary pericarp and a distinct fragrance. The pericarp forms nearly 65 per cent of the total weight of the fruit. The seeds are elongated with a thin glossy brown seed coat, and are 1.5 to 2.0 cm long containing almond-shaped whitish edible kernels. The seeds generally vary in size and weight from 450 to 1,000 per kg.

Seeds are recalcitrant in nature and lose viability within one week.

Propagation

Commonly, cheura is propagated by seed, however, hard wood cuttings under high humidity (above 85 per cent) and temperatures of 25° to 28°C, if placed in a loose and porous growing medium, also respond to good rooting. The seeds are either sown in nursery beds or in polythene bags immediately after collection. The seed should not be removed from the flesh until just before sowing, and all flesh should be cleaned off to obtain a high germination percentage. Germination takes place after 10 to 15 days under normal growing conditions. After 12 to 15 months the seedlings can be transplanted at permanent sites. The seeds are sown in June-August.

Climatic and Soil Requirements

Cheura requires a fairly warmer climate and is adapted to lower elevations (900-1,200 m) and moisture-rich sites. Frost occurrence and frequent variations in temperature during the blooming period are injuries to flowers and newly set fruits. Being a rapid grower it requires fertile, well-drained and light-textured soil with the capacity to retain adequate moisture and permitting good aeration.

Moist and shady sites with sandy loam soils are more suited to its cultivation than dry, heavy and bright sunny sites. It is a fact that its natural stands are restricted to a few specific sites and the area under this crop could not be extended for want of the peculiar climatic and soil conditions.

Planting

Twelve to fifteen-month-old seedlings can be planted in the sites offering most conducive agro-climatic conditions for its proper growth. The seedlings are transplanted at a distance of 10m apart both plant to plant and row to row.

After-care

For the proper growth of the plants the weeds are to be controlled from time to time. Well-rooted FYM and 200 gm of NPK mixture (15:15:15) per year of age of tree should be applied during October-November. If the drought conditions prevail for a longer time, arrangements for watering are essential at an interval of 7 to 10 days.

Uses

The tree is of multiple use and every part of it is used in one way or an other. Like the importance of *Madhuca longifolia* (Mahua) for the people of sub-tropical regions, cheura is of great value to the hill folk. The fruit is eaten fresh. The kernel is widely used for edible oil extraction.

13

MACADAMIA NUT (Macadamia integrifolia Maiden & Betche)

MACADAMIA NUT commonly called is Queensland nut is a member of Proteaceae family, and is supposed to have originated in Queensland and New South Wales (Australia). It is grown extensively in the upland areas of Hawaii. The tree is slow-growing and attains a height of about 10 m at full-grown stage. It thrives well in the sites which are free from frost during winters and springs and the mean temperature does not go beyond 70°C. The sites which are ideally suited to oranges and other sub-tropical ever-green species also seem suitable for its successful cultivation. Ikaika, Wailua (extensively grown in Hawaii), Keauhou and Kakea (widely cultivated in California) are the promising macadamia cultivars. The sites having sandy loam light soils and a long dry season favour its cultivation. For proper growth the soil must be provided with proper drainage facilities as water-logged conditions cause injury to the tender roots and collar zones. The tree should be protected from strong winds, therefore, the orchards should be placed either on the windless sites or protected through the dense canopied wind breaks. The soils must be fertile with a pH range of 4.5 to 6.5. The rainfall of the zone should not exceed 1,200 mm.

Propagation

Macadamia can be propagated by seeds or vegetatively by side wedge grafting techniques. The seedlings of *Macadamia tetraphylla* which is a rough-shelled nut are used as rootstock. The germination percentage varies from strain to strain in accordance with the thickness of the shell. Thin-shelled nuts germinate faster as compared to thick-shelled nuts. Seed nuts should always be husked prior to planting as it has been observed that unhusked nuts take longer to germinate. Besides, they may cause fungal rotting resulting in poor germination. Seedlings for rootstock may be raised in the nursery beds or directly at the planting site. Direct seeding method commonly known as "drop seed method" exhibits a faster seedling growth than that of the similar age transplanted from the nursery beds. The seedlings can be transplanted at 10×10 m distance in a well-prepared soil.

Cultural Practices

At the initial stages of growth up to 4-5 years, the lower branches growing around the trunk up to a height of 70 cm are removed and the lowermost scaffold limb is selected above 70 cm. The tree allows to develop a heavy top with strong evenly distributed main limbs. From time to time, the weak, diseased or overcrowded branches should be removed.

For an adult bearing tree about 2 kg of NPK mixture (15:15:15) should be applied annually along with 50 kg of well-decomposed compost. Pest management for controlling insect-like leaf defoliaters, thrips, scales, husk borers and fungal diseases like root rot fungus and cankers should be adopted well in time.

Yield

A full mature tree produces about 16-25 kg of nuts. The nuts are ready for harvesting during September-October.

Uses

The nut husk is used as animal feed after pulverizing and mixing it with molasses and pineapple bran. The hull powder is also used as filler for plastics. The nut is used in various forms for human consumption like salted, chocolate-coated nut candy and nut ice-cream. The nut contains about 60-70 per cent oil which is very rich in flavour and is widely used for industrial purposes.

14

MINOR NUTS AND MULTI-PURPOSE NUTS SUITABLE FOR THE HIMALAYAS

TABLE 14.1 gives the common name, botanical name, origin, growing conditions and uses of 23 other minor nuts and/or multi-purpose nuts. Their suitability for one part or the other of the Himalayan region needs further research and promotion depending on the findings of applied research.

TABLE 14.1: Some neglected multi-purpose nuts suitable for the Himalayas

Sl. No.	Common name	Botanical name	Origin	Growing conditions	Uses
1	2	3	4	5	6
1.	Java Almond	*Canarium commune*	Philippines	Warmer and humid sites with an elevation of 500-600 m above sea level	Nuts or seeds are edible like almonds. Seeds are used for obtaining cooking oil
2.	Country Almond	*Terminalia belerica*	Southern Asia	Lower sub-tropical elevations up to 800 m above SEA level	Edible seeds of high medicinal value
3.	Himalayan Filbert	*Corylus ferox*	Himalayan mountains	Higher temperate zone between 1800-2500m above sea level	Edible nuts
4.	American Beech	*Fagus grandifolia*	Northern Hemisphere	Higher temperate zones located between 2000-2500 m above sea level having fairly dry sandy and loamy soils	Nuts are used for feeding domestic fowls. Nuts contain 17-20 per cent of a non-drying oil suitable for lighting, cooking or as a substitute for salad oil and butter

(Contd.)

(Contd.)

1	2	3	4	5	6
5.	European Beech	*Fagus sylvatica*	-do-	-do-	-do-
6.	Alabama chinquapin	*Castanopsis alabamensis*	South-eastern United States	Suitable for warmer climatic zones	Provides edible nuts and timber
7.	Horse chestnut	*Aesculus hippocastanum*	Europe	Sub-tropical to temperate climatic zones	Starch for industrial purposes may be obtained from nuts. Nuts are not eaten
8.	Arrut or Earth chestnut	*Bunium* species	Western Europe to the Caucasus	Suitable for temperate zones	Source of feeding stuff
9.	Chilehaze nut	*Gevuina avellana*	South-western Southern America	Fairly warmer sites	Used for feeding purposes
10.	Chufa nut	*Cyperus esculentus*	Southern Europe	Lower elevations with fairly warm conditions	A useful food for pigs
11.	Souari nut	*Caryocar nuciferum*	Guyana and Brazil	Warmer and moist valleys with sandy loam soils	Fruit contains five seeds. Nuts have excellent flavour and about 60 per cent fat content

(Contd.)

(Contd.)

1	2	3	4	5	6
12.	Chir pine nut	*Pinus roxburghii*	Himalayan mountains	Fairly warmer sites with mineral acid soils between elevation range of 1000-2000 m above mean sea level.	Cone bears edible nut timber is used for ordinary furniture, resin is produced
13.	Brazil nut	*Bartholletia excelsa*	Basins of the Amazon and Orinoco rivers of South America	Warm, moist and frost-free valleys having deep light soils	Brazil nut contains 14 per cent protein and about 65 per cent fat
14.	Cathay hickory nut	*Carya cathayensis*	Native of China	Fairly warmer middle Himalayan tracts with an average annual rainfall and frost-free sites	Source of edible nuts and timber
15.	Mockernut hickory or Big bud hickory nut or Bull nut	*Carya tomentosa*	Eastern United States	Lower altitudes up to 1200 m above mean sea level. Requires adequate moisture for growth and fruit-setting. Sites should be free from frost and high rainfall with a soil pH range of 5.8-6.5	Source of edible nuts and timber

(Contd.)

(Contd.)

1	2	3	4	5	6
16.	Nutmeg hickory nut	*Carya myristicaeformis*	South-western United States	Warm and moist hill tracts located between 1000-1200 m	Edible nuts and timber
17.	Hog nut or Pig hickory nut	*Carya glabra*	Eastern United States and Ontario Canada	Middle Himalayas situated between 1000-1200 m above mean sea level having sandy loam soils	Nuts are used as animal feed and timber
18.	Shagbark Hickory nut	*Carya ovata*	South-east Canada and eastern United States	Climatic requirements are exactly as required by Hog nut	Source of edible nuts and timber
19.	Swamp hickory nut	*Carya leiodermis*	Native to Louisiana and Mississipi area of United States	Hot and moist Himalayan tracts located between 1000-1400 m above mean sea level	Edible nuts and timber
20.	Tonkin hickory	*Carya tonkinensis*	Native of southern China and North Vietnam	Fairly warmer, moist and frost-free sites with very short winters and long summers with bright and long sunshine duration	Edible nuts and timber

(Contd.)

(Contd.)

1	2	3	4	5	6
21.	Cahoun nut	*Attalea cohune*	Central America	Ideally suited to middle Himalayas situated between 1200-1500 m above mean sea level, grown better on rich, well-drained light soils	The nut yields an oil for making soap and margarine
22.	Coquita nut	*Jubaea spectabilis*	South-western South America	Lower altitudes experiencing frost-free very short winters and long summers with average rainfall not exceeding 1000 mm	The nut yields oil used for making soap
23.	Jojaba nut or Goat nut	*Simmondsia chinensis*	China, California and Mexico	Lower elevations located at 500-800 m above mean sea level with sandy loam soils	Provides foodstuff. The oil extracted from nuts is a good hair-dressing. It is also used as a substitute for whale oil.

15

MINERAL NUTRITION OF NUT CROPS

A THOROUGH acquaintance of orchardists with the mineral nutrition and fertilizer needs is of great significance for keeping the plants healthy and productive throughout their life span. Liberal or unbalanced use of fertilizers in the nut orchards can result in one element can cause the toxicity or deficiency of other elements and the trees may be subjected to many physiological disorders, in turn causing severe losses to nut yields and the longevity of trees.

Nutrient requirements of nut crops vary from area to area and species to species in accordance to soil conditions, total rainfall, crop density, topography and cultural practices followed. In spite of establishing nut orchards with potential cultivars, a grower may fail to harvest an economic yield in lack of actually demanded fertilizer programmes. The major elements like nitrogen, phosphorus, potassium, calcium, magnesium and sulphur are largely demanded essential elements but at the same time the importance of minor elements like boron, zinc, manganese, copper, etc., cannot be ignored though, these are demanded by plants in very small quantities. Under or over doses of these elements can cause different physiological disorders in the plants.

Occurrence of physiological disorders in the fruit frees due to toxicity or deficiencies of macro or micro elements is generally confused with pathogenic infectious, and lack of proper and timely treatment can seriously aggravate the problem. Therefore, proper and timely diagnosis of any disorder caused by any way use of micro or macro nutrients is of immense value for a timely tackling of the problem.

Realizing the importance of balanced use of chemicals and fertilizers this chapter dealing with the significance of essential major and minor elements and symptomatology of different physiological disorders and measures correcting nutritional deficiencies is being discussed so as to enable orchardists to diagnose the physiological disorders and to initiate suitable necessary measures for overcoming the problem.

I. **Major Nutrients**

Nitrogen

Among all the nutrients nitrogen plays a vital role in regulating physiological processes in plants and is directly responsible for controlling growth and fruiting. A shortage of nitrogen results in the reduction of the amount of chlorophyll in the leaf and hence a loss of green colour. If nitrogen is not added to soils, almost all kinds of soils sooner or later become deficient in nitrogen. The nitrogen cycle in the soil is too complex and is regulated by many factors. Largely, the supply of nitrogen in the soil occurs in three forms :

1. Nitrate nitrogen
2. Ammonia nitrogen
3. Organic nitrogenous compounds.

Nitrate nitrogen moves with the water in the soil. Ammonia nitrogen is fixed on the clay particles for a short period till it is converted into nitrate form. After converting into nitrate form it moves with the soil water. The time required by ammonia nitrogen to convert into nitrate nitrogen can be increased by addition of ammonia fertilizer of small quantities of 2-chloro-6 (trichloromethyl) pyridine. The nitrogen in the organic compounds is slowly released by the activity of soil micro organisms.

Deficiency of this element causes loss of green colour in foliage, poor shoot growth and premature and untimely defoliation. In the case of an acute shortage of nitrogen in the soil, plant growth is inhibited completely which in turn affects fruit yield adversely. Nitrogen deficiency most commonly occurs either in sandy soils under high rainfall conditions or in soils having low levels of organic matter. Contrary to this, excessive nitrogen commonly produces vigorous plants with reduced yields.

In nut production, nitrogen deficiency is the most common deficiency and the easiest to correct. For correcting nitrogen deficiency in nut orchards application of nitrogenous fertilizers may be made either through soil application or in the form of foliar spray. Nitrogenous fertilizers can be applied to the soil through broadcasting. The quantity of nitrogenous fertilizer will depend upon the nitrogen status of soil, soil types, age of plant, tree species and prevailing crop management system. Urea is widely used as the source of nitrogen for foliar application. For mature bearing nut trees, foliar application of 2 per cent urea is an ideal concentration. Growing of leguminous cover crops, use of manures and mulches in nut orchards also improve the nitrogen status of the soils.

Phosphorus

For nut crops phosphorus application is of immense value. Phosphorus is found in an inorganic state called orthophosphate. The organic phosphates are mainly phosphorylated sugars and alcohols. Phospholipids like lecithin and nucleotides are also important phosphorus containing plant organic compounds. The principal role of phosphate is to regulate metabolic processes in the plant system. In nut kernels phosphorus is stored in the form of calcium and magnesium salts of phytic acid, which is mobilized during germination. Since phosphorus is linked with various types of plant metabolites, low levels of phosphorus in the soil or an underdose application of phosphatic fertilizers hinders the metabolic processes and produces plant disorders, like spindly form of trees, poor lateral growth, yellowing of leaves, premature defoliation, reduction in blooming and delay in bud opening.

Phosphorus-deficient tree crops are subjected to slow growth and show spares foliage distribution, which is generally observed with dull bronzish to purplish tinged colour. Phosphorus deficiency is commonly found in the soils subjected to intense weathering with acidic properties or the soils containing a low level of available phosphorus. Despite the fact that these types of soil contain very high levels of phosphorus, peat and muck soils also show phosphorus deficiency. Lack of proper soil management practices like reduced soil moisture and low soil temperature affect phosphorus availability. Acid soils also restrict the availability of phosphorus to the plants. The soils containing low levels of organic material also cause phosphorus deficiency.

For controlling phosphorus deficiency, methods of

application of phosphatic fertilizers vary greatly depending upon the soil conditions, species grown and adopted crop management. Phosphorus deficiency in nut crops can be corrected by applying phosphatic fertilizers in the soil in accordance with the soil and plant fertility status, soil reaction and moisture availability. For acidic soils use of rock phosphate has been found beneficial. Phosphatic fertilizers should be applied through the band placement method by making a 50-cm-long and 50-cm-wide trench around the tree basin (basin will vary according to the actual spread of plant).

After placement, the trench should be filled with the soil. The amount of fertilizer to be applied will depend upon the nut species and the age of the trees. Lime application in acidic soils has been found effective in increasing the phosphorus availability. Use of adequate manure and organic materials, maintaining proper soil moisture and temperature also help in mobilizing phosphorus.

Potassium

Potassium has a vital role in regulating many physiological processes in the plants. It is an extremely mobile element, which moves readily towards meristematic tissues. Potassium is closely associated with growth rate, photosynthesis and cytokinin supply, enzyme activation and pigment in plants.

Potassium deficiency is commonly observed in light sandy soils, organic soils (peat and muck soils), and degraded soils. The deficiency is also common in the soils that fix potassium into non-replaceable form. The deficiency symptoms appear in the form of leaf scorching on margins, chlorosis, decline of cells, and dull colour of fruits. Under severe deficiency conditions, the plants are attacked by adverse environmental factors and diseases and insect-pests. The symptoms of the potassium deficiency appear first on the matured leaves of the plants. The chlorosis symptoms on the leaves develop from light yellow to tan becoming brown and finally drying to a scorch. Sometimes dieback of young shoots is also observed in the case of acute shortage of potassium.

Soil application of potash in the form of potassium chloride or potassium sulphate is the most common method of correcting potassium deficiency. The doses will depend upon the soil and plant nutrient status, age of the plant and species.

Organic manures are also good sources of potassium and can be applied in soil. In sandy soil where leaching is a problem much smaller amounts of potasic fertilizer are applied at short intervals. Foliar spray containing adequate but not injurious amounts of potassium salts have also been found effective. The basic objective of potassium application is to apply adequate material within the root zone, so as to increase immediate availability to the plants as well as to prevent undue losses caused by unavailable soil fixation in heavy soils or leaching from sandy soils.

Calcium

Majority of temperate nuts growing on moderate to strong acid soils are severaly affected by calcium deficiency. In walnut, pecan, chestnut, filberts, calcium deficiency symptoms are more pronounced. Acid soils having a pH below 5.5 also cause decreased rate of nitrification which in turn adversely affects plant growth. Low-calcium soil conditions result in poor availability of phosphorus, boron, copper and zinc and cause severe losses to nut orchards.

Walnut, pecan nut, hazelnut and chestnut trees growing on calcium-deficient soils exhibit physiological disorders like reduction in leaf size. Calcium-deficient nut trees produce small and yellowish leaves which show mottling and bronzing. Trees growing on calcium-deficient soils are generally damaged by rotting of root, collar and bark.

Correction of calcium deficiency is commonly accomplished by additions of lime in one form or another. Liming performs other functions as well, and responses to lime may be owing to correction of toxicity of elements like aluminium and manganese. At the same time liming also improves phosphorus, boron, molybdenum availability and nitrification rates. In calcium-deficient soils the rich sources of calcium like calcium nitrate and calcium chloride can be successfully employed. Application of organic manures also helps in correcting calcium deficiency. Calcium silicate stags from various sources will also supply available calcium when used in appropriate amounts. Exact quantities of calcium-correcting sources can be worked out on the basis of soil pH. Calcium sources selected for correcting deficiency should be well mixed into soil. For determining calcium requirements of trees, orchardists are advised to seek the help of soil and plant analysis laboratories. Alkali or sodic soils having high pH and exchangeable sodium levels are also found deficient in calcium because the plants growing on such soils often fail to secure adequate calcium due to extreme insolubility and low supply of calcium compounds. Calcium deficiency in such soils may be corrected by using gypsum. Application of 36 to 50 tonnes of organic manures or compost per hectare will also help to correct calcium deficiency.

Magnesium

Magnesium deficiency is mostly found in pecan, filbert and walnut trees growing in the areas experiencing moderate to high rainfall. Deficiency symptoms can easily be seen on trees growing on poorly drained acidic or sandy soils.

Trees suffering from magnesium deficiency show marginal yellowing of leaves with interveinal chlorosis. Necrosis can easily be seen on leaf margins and between veins. Such plants exhibit premature defoliation.

Soil applications and foliage sprays of magnesium chloride are two major methods for controlling magnesium deficiency. Foliage sprays of 2 per cent Epsom salts applied five times a year for five years during May-July can greatly reduce the symptoms of magnesium deficiency. Trees on heavy soils respond quickly to soil application of magnesium sulphate and use of magnesium sulphate containing 16 per cent magnesium oxids has been found effective in correcting magnesium deficiency.

Sulphur

Sulphur deficiency symptoms resembles the symptoms of nitrogen deficiency like reduction of leaf size, chlorosis of leaves and anthocyanin pigmentation. The only difference distinguish each other is the appearance of deficiency symptoms appear first in the older leaves, while on contrary sulphur deficiency symptoms appear first on younger leaves. Shoots are more affected by sulphur deficiency than roots. The deficiency can be corrected by application of elemental sulphur or gypsum. However, the use of elemental sulphur or gypsum will greatly depend on the soil reaction. In neutral or acidic soils, elemental sulphur increases loss of lime through leaching, while gypsum on the other hand, is a neutral salt and contains about 19 per cent of sulphur and 23 per cent of calcium.

II. Minor Elements

Manganese

Manganese deficiency generally occurs in alluvial and marshy soils which are desired from calcarious materials like calcareous silts and clays. Strongly acid soils may also pose manganese deficiency due to its leaching.

Deficiency of manganese in plants causes a physiological disorder known as "little leaf" or "mouse ear" in which leaves are deformed. In pecan nut and walnut trees such leaf disorders are most commonly found. In apricots the symptoms of manganese deficiency are exhibited by more pronounced inter-veinal chlorosis. Foliar application of 1.5 kg of manganese sulphate and 1.0 kg of soda ash per 500 litres of water helps in correcting manganese deficiency. For correcting deficiency of manganese in walnut trees, foliar application of 2 kg of manganese sulphate in the month of June has been found quite effective. In pecan, mouse ear or little leaf symptoms can readily be seen. Affected leaves are considerably reduced in size, and the tip of the mid-vein often protrudes beyond the leaf margin which in turn inhibits the growth of nuts. Depending upon age, tree size and deficiency severity, foliar application of 1 to 2 per cent manganese sulphate or soil application of 1 to 2 kg manganese sulphate can be of great help in rapid recovery of the manganese element.

Boron

Boron deficiency is very common in the soils which are derived from acid igneous rocks or from fresh-water sedimentary deposits. Acid soils are also found deficient in boron content because the boron content is removed from the soil through leaching. Likewise, plants growing on light-textured sandy soils or acid peat or muck soils suffer from boron deficiency. Alkaline soils especially those containing free lime are subjected to boron deficiency in common. Irrigated lands containing low boron in water where salt or carbonate deposition occurs also depict deficiency symptoms on plants growing on them.

Apical shoots of boron-deficient almond trees start drying from top downwards but lateral shoots farther down may grow, giving a brushy appearance. Leaves of such plants show curling and premature shedding and external surface of the fruits exhibit gummy lesions. Such deficient trees produce plumpy kernels. Premature nut drop oftenly occurs in May-June. Boron deficient leaves of walnut, pecan and hazelnut trees show dark brown spots between veins and the tip of shoots starts drying and cause defoliation. Boron deficiencies can be visualized in apricot as drying of terminal twigs. Leaves of deficient apricot plants are spatulate, dwarfed and curled at margins. Such leaves are narrow, brittle and chlorotic between veins. Fruits develop brownish lesions in flesh which is more prominent near the stone.

Boron deficiency can be controlled by Borax applying 250 gm of per bearing tree through soil application. In the case of walnut, pecan and chestnut the doses of borax for mature bearing trees should be increased up to 1 to 1.5 kg per tree. These doses of borax would be adequate for three-four years. Soil treatment through borax application should be done in late winters or early spring. Foliar spray of 0.5 per cent boric acid (1 kg boric acid dissolved in 500 litres of water) has also been found effective in overcoming boron deficiencies in nuts. In case borax is not available for soil application, foliar application of boric acid may be done three weeks after flowering which will take place between late March and the first fortnight of April.

Zinc

By and large, zinc deficiency is commonly found in almost all types of soils but it is most severe in acidic, sandy soils, alkaline soils, soil derived from granites, gneisses and soils containing clays with low Si/Mg ratio. Excess and prolonged use of phosphatic fertilizer have also been observed as a limitation on zinc uptake. Nitrogen fertilization often increases the severity of zinc deficiency. Zinc-deficient nut trees often show mottling and chlorosis of leaves and produce numerous smaller leaves. The physiological disorder caused by zinc deficiency is referred to as "Rosette". Leaves and leaflets of zinc-deficient trees appear yellowish, narrow and crinkled in the advanced stages of the disorder. New shoot growth is inhibited, internodes are short and the foliage appears bunched or rosetted. In the chronic stage, shoots die back from the tip. This die-back is usually confined to the current year's growth. Affected trees bear poorly filled nuts or blind nuts. Depending upon the species, tree age and vigour zinc sulphate can be applied the soil or as foliar spray. In mature trees one-two kilogram zinc sulphate per tree may be applied evenly around the outer skirt of the tree basin. For very young trees with poor branching and conopy, about 50-100 gm of zinc sulphate per tree will be sufficient for soil application.

Rosette may be corrected by direct sprays on the foliage. Foliar spray has been found more effective than that of soil application. For preparing ideal concentrations of zinc, dissolve two kilograms of zinc sulphate and one kilogram of slaked lime in 500 litres of water. Spray applications of zinc sulphate should be done between May and June.

Copper

Copper is involved actively in protein and carbohydrate metabolism. It has also been found that copper has a symbiotic relationship in nitrogen fixation. Copper deficiency has been found very common in deciduous trees. Deficiency symptoms at the early stages are difficult to detect but no occurrence of moderate to acute shortage, the terminal growth of the plant ceases and results in the die-back of growing twigs, rosetting or shortening of internodes of terminal leaves. Terminal leaves may or may not show physiological disorders like chlorosis, necrotic spotting, dull appearance of leaves and reddish browning of veins. Copper-deficient plants generally produce 'S'-shaped shoots. Higher doses of nitrogen application can also cause copper deficiency. Similarly, phosphorus accumulation in soils resulting from prolonged use of phosphoric fertilizers will produce copper deficiency in some soils. It causes summer die-back which is commonly referred to as "Exanthema". At the initial stage of the disorder, the leaves are twisted and the plant produces vigorous water shoots.

For correcting copper deficiency in deciduous fruit trees copper sulphate ($CuSO_4 .5H_2 0$) commonly known as blue vitriol at the rate of 0.3 per cent concentration can be sprayed on the foliage, to avoid phytotoxicity caused by sulphate ions. The solution should be neutralized by adding decanted solution of slaked lime (2 parts copper sulphate and 1 part slaked lime).

Molybdenum

The major function of molybdenum in plants is in nitrogen metabolism. It is the major constituent of nitrate, reductase and nitrogenase, which are required by rhizobia for nitrogen fixation. The deficiency symptoms appear on the lower leaves in the form of chlorotic inter-veinal mottling, marginal necrosis and infolding of the leaves.

Molybdenum deficiency symptoms can be corrected through foliar spray of sodium or ammonium molybdate (one gm is to be dissolved in 500 litres of water). Thorough coverage during spray is essential to get the best results. In acid soils, where total molybdenum is adequate but available molybdenum is low as a result of low pH, molybdenum deficiency can be controlled by lime application. The quantities of molybdenum salts in the form of sodium or ammonium molybdate should be used only after thorough consultation with the soil specialist as excessive application may cause hazardous effects on the ruminants consuming grasses and fodder crops of that area.

16

INSECT-PEST AND DISEASE MANAGEMENT IN MAJOR TEMPERATE NUTS

WITH the increase in area under fruit crops, various destructive agencies, lowering both yield and quality of fruits and other tree products, have also increased enormously. Proper recognition of these destructive agencies is the prime step in diagnosing the cause of problem as well as in ascertaining its effective control measures. Like other fruit crops, the deciduous nuts are also damaged by agencies like climatic hazards, insect-pest and diseases. The diversity in their types, mode of infection/infestation and degree of severity considerably vary from place to place and crop to crop, and are very difficult to understand, until one has a thorough knowledge about the causes of damage.

Nut crops are damaged by a number of diseases and insect-pest. Some of them are more prevalent in the area. Therefore, to obtain desirable yield and quality of nuts, it becomes quite imperative to follow suitable and timely pest management systems. This chapter dealing with the major insect-pest and diseases is meant to serve as ready reference for the nut growers to enable them in detecting the real cause of damage on the basis of the mode of infestation/infection and spread of disease, and taking effective timely action to safeguard the yield and prevent quality losses.

TABLE 16.1: Major insect-pests of temperate nuts and their control measures

Common and scientific name of insect-pests	Nut crop damaged by insect-pests	Mode of Infestation	Control Measures
1	2	3	4
Peach leaf curl aphid (Brachycaudus helichrysi)	Almond	Aphids suck the sap from leaves, vegetative buds and floral buds. Infested leaves acquire curling and turn light greenish to yellowish in colour. Plant bears poor fruit setting. Premature fruit drop occurs. High degree of incidence is noticed at pre-bloom and blooming period. Beside damaging almonds, other stone fruit like peaches, plums and prunes are also infested by this pest.	Methyl demeton @ 0.025% should be sprayed before 7-10 days prior to floral bud opening, i.e. at pink bud stage. or Apply dimethoate @ 0.03% or monocrotophos @ 0.04% or fenitrothion 0.05% in the form of foliar spray. For obtaining good results give a thorough spray on plant surface particularly on bud axils.

(Contd.)

(Contd.)

1	2	3	4
Blossom thrips (Taeniothrips rhopalantennalis)	Apricot and Almond	Thrips damage the floral buds. Magnitude of damage is severe at pre-bloom stage.	Pre-bloom spray at pink bud stage of insecticide like methyl demeton @ 0.025% or monocrotophos 0.04% has been found most effective in controlling blossom thrips.
Stem borer (Aeolesthes holosericea) Shothole borer (Apriona cinerea)	Apricot and Almond	The infestation can easily be seen from the faecal pellets as well as from drying up of the terminal shoots. The attack of stem borer is usually seen on the main trunk at 10-45 cm above ground level. Due to shot-hole borers, symptoms of damage are apparent as shot-hole like perforations on the main trunks and limbs.	For controlling stem borers, clear the hole with flexible wire and then insert the cotton wick soaked in petrol or methyl parathion (0.2%) or insert 0.5% paradi-chlorobenzene (PDCB) and plug the hole with wet clay soil. In the case of localized infestation caused by shot-hole borer, treat the affected part with methyl parathion 0.2%. In the case of severe infestation spraying with fenthion 0.05% may work satisfactorily under the initial stages of infestation.

(Contd.)

(Contd.)

1	2	3	4
San Jose Scale (Quadraspidiotus perniciosus)	Almond	San Jose scale such the sap from stem, branch, shoots, leaves and fruits of adult trees and young nursery stock. Due to continuous sucking of sap by the insect the plants lose vigour and inhibit plant growth. The heavily infested nursery and young plants die.	It can effectively be controlled by spraying the trees with diesal oil emulsion made from 5 litres emulsified with one kg. of ordinary soap, 750 gms of potash vegetable oil soap and 70 litres of water. The spraying should be done at dormant stage during December-January. For controlling damage of San Jose Scale in nursery, treatment of nursery plants with 3% ESSO tree oil followed by spraying of 0.06 to 0.1% parathion day before digging them for out-planting should be done.
Bark-eating beetles (Aeolesthes sarta, Sthenias grisator and Sphenoptera lafertei)	Almond, apricot, chestnut, hazelnut	The beetles cause a mild damage to stem and tender limbs by boring inside the bark or by girdling it.	After removal of webbings treat the affected parts in March, September and October with methyl parathion 0.2%.

(Contd.)

(Contd.)

1	2	3	4
Leaf-eating beetles (Adoretus nitidus, Myllocerus discolor, Amblyrrhinus porricollis)	Almond, pecan nut, hazelnut	Cause considerable defoliation resulting in poor yields.	Spraying of Endosulfan @ 0.15% has been found most effective. In Almonds spraying should be done either prior to floral bud opening or at husk-off stage.
Tent cater pillar (Malacosoma indica, Parasa repund)	Almond, chestnut, hazelnut, pecan nut, apricot	Caterpillars feed on leaves causing defoliation.	Spraying the trees with carbaryl 0.1%/Malthion 0.05%/endosulphan 0.07% is effective for the control of caterpillars. Spraying should be done in the afternoon with an interval of 10-12 days. In all 2-3 sprays are required.
Walnut aphis (Chromaphis juglandica)	Walnut, pecan nut, and occasionally on chestnut.	The aphids owing to pale green colour are difficult to spot owing to their location along the midrib on the underside of the green leaves. The heavily infested plants appear sickly.	Spraying with Methyl-demeton 0.025%/monocrotophos 0.036% is quite effective for the control of aphids. Two to three sprays should be given at an interval of 10-12 days.

(Contd.)

(Contd.)

1	2	3	4
Defoliating beetle (Monolepta erythrecephale, Lachnosterna corriacea)	Walnut, pecan nut, chestnut, hazelnut	Beetles defoliate the plants. As a result of high magnitude of damage the plants lose their vigour and fruit yields.	Spraying with carbaryl 0.1% or monocrotophos 0.5% is effective for the control of this pest. The first spraying should be given after first rains in the month of May-June when the beetless emerge out in large numbers.
Walnut weevil (Alcidodes porrectirostris)	Walnut	The adult weevils feed on leaves, petioles, shoots, buds and fruits. The grub bore inside the fruit and feed on the kernel turning it into a black mass. The damaged fruits fall on ground.	The pest may be easily controlled by timely picking of fallen fruit and their burning. The trees should be dusted with 5 to 10% BHC once or twice to kill at the hibernating beetles on emergence.

(Contd.)

(Contd.)

1	2	3	4
Indian mealmoth (Plodia interpunctella)	Pistachio nuts and other dried fruits	Larvae of the Indian meal moth causes severe damage to stored nuts.	Infested nuts should be culled prior to storage.
Filbert bud mite (Phytocoptella avellanae)	Hazel nuts	Tender vegetative and floral buds are damaged severely resulting in poor nut yields.	2-3 sprays of Dicofol 0.1% wettable, sulphur 0.2% are quite effective against the mite.
Pecan weevil (Balaminus caryae)	Pecan nut	Mode of infestation is same as given in walnut.	Control measures given for controlling walnut weevil may be applied for controlling pecan weevil.
Codling moth (Laspeyresia pomonella)	Walnut and pecan fruit	Kernel of the fruit is damaged by the larvae of this insect which is pinkish brown in colour.	Spray of endosulfan 0.15% should be sprayed during May-June. Spraying should be repeated thrice at an interval of 10 days.

TABLE 16.2: Major diseases of temperate nuts and their control measures

Name of disease and its causal organism	Symptoms	Control measures
1	2	3
Almond Brown rot, blossom blight (Monilinia laxa)	Young blossom spurs and associated leaves collapse to form shoot blight. Gum may extrude at the base of infected flowers, cankers on blighted twigs have tan centres with dark margins. At high humidity, gray to tan spore masses form on diseased flower parts and twigs cancer. Flowers may become infected from pink bud to petal fall and are most susceptible when fully open. In almonds, stigma, anthers and petals are all very susceptible to infection. The fungus survives in twig cankers and on remaining diseased flower parts and spurs. Spore pads (sporodachia) appear on these structures in late winter to provide inoculum for spring infection. Spores are airborne or rain splashed. Infection is favoured by rainy weather.	Spray tree with Difolatan @ 0.15% or Captan @ 0.3% at pink bud stage. Repeat it after 15 days at husk-off stage. Under severe infection of blossom wilt or blight spray carbendazim fungicide @ 0.05% at first flower and repeat it after 10 days.

(Contd.)

(Contd.)

1	2	3
White root rot (Dematophora necatrix)	Affected trees show sparse foliage, slow growth, bronzing and yellowing of leaves. Such trees ultimately die. Roots turn brown and remain covered with white cottony mycelial growth of the fungus in rainy season.	Improve the drainage of the orchard. Remove the infected roots and apply chaubattia paste (Red lead oxide, copper carbonate and linseed oil in 1:1:1.25 ratio respectively) on the cut ends and healthy portion of the rots during November-December. For curing the ailing tree, drench the soil at 15-30 cm depth prior to onset of monsoon by 0.1% carbendazim. Repeat the treatment thrice at an interval of 15 days.
Collar rot (Phytophthora cactorum and P. citrophthora)	Collar region near ground level turns brown soft and spongy and on being girdled completely kills the tree. Disease spread is maximum in heavy and poorly drained soils.	During November-December remove the soil around the collar region and expose the affected portion to sun. Remove the affected bark and apply chaubattia paste or copper oxychloride paint on the healthy portion. Irrigate 30 cm radius around

(Contd.)

(*Contd.*)

1	2	3
		tree trunk with Dithane M-45 @ 0.4% or copper oxychloride @ 0.5%. Avoid water logging in the orchard.
Shot hole disease (Stigmina carpophila)	Spots occur on leaves, fruits, twigs and flowers. However, flower and twig lesions are relatively scarce or difficult to find. Leaf lesions begin as tiny reddish specks that enlarge by several millimetres into spots having tan centres and purplish margins. When the fungus sporulates, a small dark speck appears in the centre of the spot which is a diagnostic characteristic of shot hole disease. Spots on young leaves usually fall out, leaving a hole (the shot-hole). Older leaves retain their lesions. Fruit spots are small with purplis margins, slightly corky and raised. Spots are found on the upper surface of fruit with respect to the way it hangs on trees. Heavy infection may cause distortion or gumming of fruit.	Spray with copper oxychloride @ 0.3% or Mancozeb @ 0.25% before rainy season.

(*Contd.*)

(Contd.)

1	2	3
Leaf blight (Hendersonia rubi)	Leaves at the base of shoots wither and die in early summer. The dead area extends to the base of the petioles. Leaf blades disintegrate, but diseased petioles stick on tree during winter. The fungus survives on dead petioles. Spores are spread by rain, and disease is favoured by wet spring weather. Leaf blight destroys more than 20 per cent of the leaves in a season. However, repeated early death of leaves may result in weakening of trees and poor yields.	Spray captan or Ziram @ 0.3% during the period from bloom through early spring. Apply first application during early leafing. Repeat spray at full bloom and petal fall stages.
Almond scab (Cladosporium carpophilum)	Greyish-black, soft looking spots form on leaves, fruit and twigs. Young lesions are indistinct yellow specks, best seen by holding a leaf up to the light. Lesions usually are not visible until late spring to early summer.	Spray Captan @ 0.3% at pre-bloom and husk off stage.
Powdery mildew (Podosphaera tridactyla)	White milky powder appears on the young leaves and current season's growth. Further growth of infected buds, shoots and leaves is checked.	Spray wettable sulphur @ 0.3% or carbendazim @ 0.05% at pink bud and at husk off stages.

(Contd.)

(*Contd.*)

1	2	3
Almond rust (Pranzocheschelia discolor)	Rusty pustules appear on the affected leaves.	Spray carbendazim at the rate of 0.05% in the month of May. Repeat it at an interval of 15 days till first fortnight of July.
Crown gall (Agrobacterium tumefaciens)	Rough, abnormal galls form on roots or trunk. Galls are soft and spongy. The centres of older galls decay. Young trees become stunted, older trees often develop secondary wood rots. The bacteria survive in gall tissue and in the soil. Bacteria enter through wounds. Crown gall is most damaging to young trees and nursery stocks.	Prior to establishing almond orchards sterilize the soil with formaldehyde solution. Treat the infected trunk, limbs and foliage and roots by spraying streptocylin @ 0.01% + copper oxychloride 0.2% at 15 days' interval. Also trench the soil with the solution to control infected roots. Existing galls should be removed.
Bacterial gummosis/ bacterial shoot blight/blast (Pseudomonas syringae)	Circular to elongated soaked gumming lesions arise on bark or outer sap wood and fruit. Under severe conditions terminal die back of shoots occurs.	Spray streptocychin @ 0.03% before the onset of rains. During December spray copper oxychloride @ 0.3%.

(*Contd.*)

(Contd.)

1	2	3
Walnut Anthracnose (Gnomonia leptostyla)	Blotching on the leaves occurs, circular necrotic spots on nut husk appear and nut fall prematurely. Cankerous symptoms appear on twigs.	Spray Bordeaux mixture (4:4:50) or copper oxychloride @ 0.25% or captan @ 0.3% at leaf unfolding stage. Repeat the spray operations after 15 days till leaves attain well expanded form.
Die-back (Glomerella cignulate)	Irregular dark brown spots appear on leaves which involve the entire leaf causing defoliation. Blighting of twigs is commonly observed in young plant. In the case of severe infection young plants may die.	Spray Mancozeb or captan or copper oxychloride @ 0.3% before onset of monsoon. Repeat the spray if disease prevails.
Pecan nut Scab (Fusicladium effusum)	Symptoms appear on new foliage as olive brown lesions on the ventral (lower) leaf side and later on the dorsal (upper) surface. Initially lesions are pinpoint in size but soon enlarge to 3-5 mm diameter and coalesce. These lesions on shucks are small, black, and circular and appear sunken as the nut-shuck enlarges. Surface of the	During dormant season collect the fallen leaves and burn them. Spraying of Dodine @ 0.1% or Dithianon @ 0.05% should be done before onset of monsoons. Repeat the spray operations at an interval of 10 days till August first week.

(Contd.)

(Contd.)

1	2	3
	shuck may become blackened by coalesced scab lesions. Immature affected nuts fall prematurely prior to harvesting or may fail to fill. Lesions may develop on current season leaf petioles and contribute to leaf abscission. The primary source of over-wintering innoculum is leaves, shucks and twigs.	
Downy spot (Mycosphaerella caryigena)	Lesions appear in early summer as white tufts or "frosty" spots on lower sides of leaflets. Later, in the summer the spots become greenish yellow on both leaf surfaces then chance to brown in the fall. As the season advances, diseased areas die and some premature defoliation occurs. Leaflet abscission from the leaf rachis may begin in early September. The fungus over-winters in the diseased leaves where its life cycle is completed.	Spraying of Zineb @ 0.3% onwards April till August at an interval of 15 days has been found effective in controlling the disease.

(Contd.)

(Contd.)

1	2	3
	During rainy periods in spring, new leaves are infected by fungal spores disseminated from fruiting bodies of the pathogen in old leaves.	
Leaf blotch (Gnomonia caryae)	Blotching of the leaves occurs and circular dead spots appear on the nuts. Cankerous symptoms appear on twigs.	Spray copper oxychloride or Mancozeb @ 0.3% at leaf unfolding stage and repeat it at an interval of 15 days till leaves have fully expanded.
Pistachio nut Pistachio wilt (Verticillium albo-atrum)	Wilt is caused by soil inhabiting fungi which attacks the roots and stem of pistachio plants.	Use wilt resistant cultivars. Pre-plant soil sterilization either with soil fumigants or by formaldehyde solution.
Crown rot (Phytophthora parasitica)	Rotting of collar region near ground level starts. Initially infected collars show brownish, soft and spongy appearance which at later stage girdles and kills the plants completely.	*P. terebinthus* should be used as rootstock. Drainage in the orchard should be improved. Treat the affected part with chaubattia paste after removing affected parts. Drench the soil near collar zone up to 30 cm depth with 0.4% Mancozeb or 0.5% Copper oxychloride.

(Contd.)

(Contd.)

1	2	3
Chestnut Seedling blight (Sclerotium spp.)	Infected seedling gets killed. Small mustard coloured sclerotia appear at the collar region during high humid conditions.	Destroy the affected seedlings. Irrigate the nursery soil with 0.3% Thiram solution or 0.04% Aureofungin.
Hazel nut Twig blight (Apioporthe anomala)	Drying of leaves starts from margins which at advance stages covers the entire leaf. The affected leaves fall prior attaining full size.	Spray 0.15% Difolaton on occurrence of symptoms. Repeat spray thrice after every 15 days.
Root rot (Armillaria mellea and Phymatotrichusm omnivorum)	Leaves show yellowish colour resembling nitrogen deficiency. Plant growth is checked completely. Premature leaf-fall takes place under severe infections and plants get killed.	Drench the soil with 0.4% copper oxychloride or 0.3% captan.
Chilgoza pine nut Seed rot and damping off of seedlings (Pythium spp. and Rhizoctonia spp.)	Seedling starts falling down from the collar zone just near the ground level due to killing of tissues.	Seed treatment with Captan or Thiram @ 2.5 gm/kg seed weight should be done before sowing. Nursery soil should be treated with 0.3% captan solution or with formaldehyde solution. While using formaldehyde, it must be ensured that nursery soil is left for 7 days until it becomes free from formaline odour.

17

NECESSARY PRECAUTIONS WHEN USING PESTICIDES

FOR controlling insect-pests and diseases various kinds of insecticides, rodenticides, fungicides, bactericides, nematicides, etc., are widely used by orchardists. Timely application of the desired concentrations of these chemicals, help orchardists in controlling the losses caused by insect-pests but if proper precautionary measures are not taken during their application or storage, they may prove to be quite hazardous for humans and livestock. Therefore, it is highly desirable that proper precautionary measures are adopted while spraying, broadcasting or applying pesticides in any other form. Orchardists must remember that almost every pesticide is poisonous and must be used with utmost caution. Following suggestions are, therefore, made to avoid health hazards :

1. Read the label carefully before opening a pesticide container.
2. Follow all precautions and directions given on the label on the pesticide's container including rate/concentration, method of application, date of manufacture and expiry and protective equipments needed during its use.
3. Apply pesticides at the specified rates and on the crops or in the situations as per the

instructions given on the label.

4. Do not carry or transport pesticides with food or feed.
5. Always keep pesticides in original containers until they are used. Store them in a locked cabinet, room or fenced area where they are not accessible to children, unauthorized persons, pests or livestock. Never store pesticides with foods, feed, fertilizers or other materials that may become contaminated by the pesticides.
6. Dispose of empty containers carefully. Never try to re-use them. Users should ensure that empty containers are burried properly so that they are not accessible to children or animals. Never dispose of containers where they may contaminate water supplies or natural waterways. Prior to using and disposing of the pesticides and their containers consult Plant Protection officers or Horticulturists about the correct procedures.
7. Under certain conditions some chemicals may cause injury to certain crops (phytotoxicity). The user should also ensure that the pesticide possesses compatability if more than one pesticide is used or when pesticides are used in combination with fertilizers or trace elements (micro-nutrients). If they are not compatible with each other they may be highly injurious to the crops and cause serious losses to the orchardists due to their harmful effects on plants. Therefore, confirm the compatability with the help of the compatability chart prior using any pesticides. (See the compatibilities chart of the end of this chapter).

8. Before using any pesticide considerations like stage of plant development, soil type and condition, temperature, moisture and wind direction, velocity, etc., should be paid due attention.
9. Some materials or rates cannot be used in certain crops within a specified time before harvest. Strict adherence to label instructions be ensured and required time gap between application and harvest of produce should be maintained. Negligence may cause serious health hazards.
10. Shelled or in-shell nut kernels which are selected for human consumption should not be treated with any kind of pesticides so as to safeguard human health from harmful effects of pesticides. However, the nuts to be used for raising nursery stocks may be treated with an appropriate dose and kind of pesticide to avoid losses during storage as well as to control seed-borne diseases and insects.
11. Many pesticides are toxic to pets or beneficial animals including honey bees, fish and birds. Crops and other plants may also be damaged by misapplied pesticides. Take every precaution to protect non-pest species from direct exposure to pesticides and from contamination due to drift, runoff, or residues. Certain rodenticides may cause hazard to animals that eat poisoned rodents. Left-over poison baits of rodenticides should be burried carefully in the soil so that they are beyond the reach of domestic animals.
12. Crops demanding cross pollination through insects should never be given any pesticide application just prior to opening of flowers or during flowering. Carelessness in this regard

will certainly cause serious losses to useful pollinating insects and result in low fruit set and yields due to inadequate pollination.

13. While using pesticides follow very carefully all essential protective measures as per directions given on the label of the containers. Avoid splashing, spilling, leaks, spray drift and contamination of clothing. Never eat, smoke, drink or chew while using pesticides.
14. On occurrence of any health complication due to misapplied pesticides consult a doctor immediately.

Appendix I

HIMALAYAN NUTS PROJECT – PHASE II

INTEREST has been expressed as a result of a discussion with Dr. R M Pandey and the follow-up of the work that has been initiated through voluntary agencies in the Central Himalayas with the participation of Trees for Life (TEL) (Report of the Nuts Planting Training Programme Enclosed).

It is proposed to have the following ingredients in the Phase II
project.

1.	Improved varieties of walnut	—	A total of 500,000 seedlings to be developed over a period of three years.
2.	Almond	—	100,000 seedlings
3.	Sweet Chestnut	—	100,000 seedlings
4.	Apricot	—	100,000 seedlings
5.	Chilgoza	—	50,000 seedlings
6.	Pistachio	—	50,000 seedlings
7.	Pecan Nut	—	50,000 seedlings
8.	Hazelnut	—	50,000 seedlings

A total of over one million nut seedlings and plants over a period of three years.

This will involve a number of voluntary agencies such as :

(i) Berinag Gram Swarajya Mandal - Kandey
(Shri Sadan Mishra)

(ii) Uttarakhand Jan Jagriti Sansthan Khad Jalal
(Shri Pratap Shekhar)

(iii) Parvateeya Jan Kalyan Samiti
(Shri Yogesh Chandra Bahuguna)
(iv) Chirag
(v) S.O.S. Prakriti
(Shri M.L. Joshi)

Following points need to be:

(i) Procurement of seeds.
(ii) Provision of management, development of nurseries and all other aspects to be worked out on the lines as in Phase I.
(iii) The estimated cost including seeds, transport, promotion of nurseries, arrangement for additional water through springs, etc., is equivalent to Rs. 10 lakhs over a period of three years (1991-1994).

Appendix II

DISTRIBUTION OF SEEDS FOR DEMONSTRATION

S.No.	Name of the farmer and address	Seed	Quantity
1.	Dhani Ram s/o Shri Bogia,	Walnut	5 kg
	Vill. Lotou, Block Kalsi,	Apricot	1/2 kg
	Dehra Dun.	Almond	1/2 kg
2.	Sri Gulab Singh s/o Shri Rithu,	Walnut	5 kg
	Vill. Hamron, Block Kalsi,	Apricot	1/2 kg
	Dehra Dun.	Almond	1/2 kg
3.	Demonstrations at Khadi (Walnut)		3 kg
	Hill Campus		2 kg
	Ranichauri, Jagdhar (Yogesh Bahuguna farm)		2 kg
	Sabli Mahila Mandal		3 kg
	Govt. Intermediate College (GIC) Ranichauri		2 kg
	Total walnut seed used		22 kg
4.	Allotment to Hill Campus		2.68 qtls
5.	To Mr. Sadan Mishra		5.34 qtls
6.	To Mr. Yogesh Bahuguna		3.80 qtls
7.	To Sabli Mahila Mangal, Sabli		1.90 qtls
8.	To Principal G.I.C., Ranichauri		0.63 qtls
9.	To Sri Pratap Shekhar, Khadi Jalal		7.00 qtls
			21.68 qtls or 2168 gms

Appendix III

PRESENT STATUS AND RESEARCH AND DEVELOPMENT OF PROMOTION AND PRODUCTION OF SWEET-CHESTNUT IN HIMALAYAS – A BRIEF NOTE

SWEET-Chestnut is a dry fruit which can mostly be grown from 5,000 ft. to 8,000 ft. from sea level. It is very rich and proteneous fruit and very easily digestable. It has a very big tree and survives for more than 100 years. One matured Sweet-chestnut tree can fruit up to 150 kgs to 200 kgs at a time. Since it is a broad-leaved specie, it makes a thick layer of 1" to $1^1/_2$" during autumn and maintains the humous of the ground for whole of the year. The leaves are soft and easily decompostible and very fine for organic manure sometimes better than Oak leaves. It can be used as a fodder also during spring season. It is a type of wild species and can be grown in jungle along with other types of wild species of broad leaves.

During the British regime the English people planted Sweet-chestnuts in some of the places of Kumayun hills but not in a very large scale. The local people did know its use and utility very little. Moreover, it is difficult to pick up it from the tree since it has needle-like thorns in its outer cell. But still, there are some trees of British time and they are fruiting very well.

During the last ten years we happened to know about its utility and richness in nutrition from some of the old people of British time. Since then only we could try to make some experiments on it. For the last seven years we have been trying to develop it under our NUTS DEVELOPMENT Programme as a main species. We believe under out tree culture programme for the economic benefit of the local people and the environmental protection as well, nuts development programme is only the alternative for the Himalayan region.

Dr. M.L. Dewan, personally, and his organisation TREES FOR LIFE has been kind enough to develop and promote a Nuts Development Programme in which Sweet-chestnut development is a major activity.

Every year, for the last three years, we have been growing more than 25,000 saplings of Sweet-chestnut only and distributing the same to the local people. In some of the places it has started fruiting also. We tried our level best to grow it in large numbers but seed-collection is a big problem and at the same time the germination rate is also only 40 to 50 per cent.

By now, we have been able to produce and distribute near about one lakh seedlings in whole of the Uttarakhand area of the Himalayan region. We plan to have a massive programme for its development i.e. one lakh seedlings every year to be grown and distributed amongst people.

At present it has a good market also. Mostly, it is being used by some of the confectioners and sweet shops. But very few people know about it. At present it is being sold between Rs. 16 and Rs. 20 per kg at BHOWALI, Nainital. Anyway, it has a bright future. We are committed for its promotion and development.

Berinag Gram Swarajya Mandal,
P.O. Kandey, Berinag,
Distt. Pithoragarh
October 4, 1991

SADAN MISHRA
Secretary
PIN: 262531

Appendix IV

KINNAUR'S UNIQUE CHILGOZA FOREST CAN STILL BE SAVED

KINNAUR district of Himachal Pradesh has the rare distinction of growing chilgoza famous for its edible seeds. Chilgoza, which falls in the preferential category of dry fruits, is grown under natural conditions in a sizeable wooded area of the district. It fetches a handsome price and forms a substantial source of income for tribal families.

In Kinnaur district, chilgoza pine trees grows in the dry temperate forests which remain outside the influence of monsoons. Chilgoza pine grows in the company of deodar and chir pine which beautify the otherwise dreary and rugged mountain slopes and add to the scenic charm of the area. It endures severe winter and makes little demand on the fertility of soil. It is capable of growing on very dry hills with shallow soils.

According to records available, forests once occupied a much greater area of Kinnaur. The presence of chilgoza was first noticed by the Gerard brothers in 1817 when they undertook a survey of forests of Kinnaur. After this, Royal surveyed forests of Kinnaur in 1832 followed by Thompson in 1847 and Cleghorn in 1864 and they also reported the availability of chilgoza in the thick forests of Kinnaur.

On account of marked variation in annual precipitation along the entire length of Sutlej valley, the tract of Kinnaur can be divided into three climatic zones: wet, dry and arid. The wet zone comprises of lower or outer valley extending from Maneoti dhar in the Nichar range to Kilba along the northern and upto Wangtu along the southern aspects. This zone receives the bulk of rainfall during the rainy season from the south west monsoons and heavy snowfall in winter.

The dry zone lies between Kilba and Purbani including forest of Baspa valley on the left of the Sutlej and between Wangtu and Kalpa on the right of the Sutlej. It extends over parts of Kilba and Kalpa ranges. The rainfall in this zone shows a progressive decline and snowfall during winter is heavy.

The arid zone lies beyond Purbani and Kalpa extending to the Indo-Tibetan border. This zone does not get any rainfall and winter in this zone is severe. It is in the dry temperate forests that chilgoza pine grows.

In Kinnaur, chilgoza pine replaces chir near Wangtu and continues along the Sutlej as far as Kanam. Near the bed of the Sutlej at a level of 2,133 metres and in intensely arid conditions, chilgoza pine forests persist as a low level species. At certain places, chilgoza extends right up to alpine levels.

Although Kinnaur is the only place in the country producing chilgoza, still its forests need proper management. Chilgoza forests are subjected to use by local people grazing besides collection of nuts. The chilgoza pine forests in Kinnaur are deficient in natural regeneration due to unrestricted collection of chilgoza seed by right holders who leave hardly any cone on the trees. Besides, birds, rodents and reptiles destroy the seed in otherwise inhospitable climatic conditions which hamper the establishment of seedlings.

There is at present no established technique of

raising chilgoza plantation. The Forest Department is attempting restocking of suitable areas having compact manageable sites with good soil for raising chilgoza plants. Considering the importance of chilgoza as forest produce a plan has been formulated for raising new plantation and maintenance of old chilgoza trees under a centrally sponsored scheme. Under the plan, the area is proposed to be closed for a specific period, fenced with barbed wire and watched to prevent grazing. Because of the long spell of dry weather, the plantation will be irrigated once in a fortnight for the first two years during dry periods.

Now, when chilgoza forests have assumed economic importance, its proper management is essential for maximum production for trade and regeneration. Considering the economic, social and conservation aspect of chilgoza forests, it is essential to undertake demarcation and settlement of chilgoza pine forests. It is essential to declare chilgoza as a "reserved tree".

The tribals of Kinnaur have the right to collect chilgoza from forests without any hinderance. Before independance, the bulk of the country's requirements of chilgoza were met from Baluchistan and its market price was also low. Chilgoza from Kinnaur made its commercial appearance only when the land-locked valley was opened up with roads.

Besides chilgoza, Kinnaur is considered as the home of quality almond, walnut drying type of apricot, drying type of grapes and pistachio nut.

Dry fruits of Kinnaur find a ready market outside. Grapes are used for making local wine, "angoori", whereas "kaghazi" almonds grown here match the best quality of almonds grown anywhere else. Almond growing areas in the district are Spillo, Ribba, Ropa Valley, Pooh, Kanam Telangi, Namgia,

Purbani and Rarang.

The almond varieties grown in Kinnaur district are dhebar, telangi, non pareil, ribba selection, local selection, thin-shelled and ne-plus ultra. The varieties of walnut include Govind Akhrot, Kashmir budded, Eureka, Franquette, local selection.

Grapes are grown in Sharbo, Ribba, Chholtu and Tapri while apricot is grown in Peo, Ribba, Spillo, Pooh, Chango and Purbani Walnut is grown in Peo, Shong (Powari) Kilba, Ribba, Kamru and Bhaba.

The drying type apricot varieties are Charmagaz, Suffaida, Shakar-para, Kaisai and Nari, while drying type grape varieties are Thompson. Seedless and Kali Sahibi. Famous raisin grape varieties of Afghanistan have also been introduced in the region and results are encouraging.

The Himachal Pradesh government is trying to commercialise dry fruits. It has set up progeny-cum-demonstration orchards in Kinnaur district having dry fruit plantations. These are located in Boktu, Ribba, Spillo, Gyabong and Pooh. Although the horticultural produce of the area meets the international standards and is getting a premium price in the national market, yet there is need to increase production per unit area which is low at present.

The University of Horticulture and Forestry is exploring the possibility of introducing dry fruit varieties available in Soviet Central Asia (Kazakhistan and Uzbekistan) to the dry temperate area of Kinnaur where climatic conditions are similiar.

Tribune 15.8.1991 S.K. BHALLA

Appendix V

ALMONDS – THE ULTIMATE 'NUTRINUTS'

IF you are looking for food with high nutritive value, your search ought to end with a sachet of almonds.

Why? Because almonds rank first in a nutritional ranking among nuts such as pistachios, cashewnuts, walnuts, etc. They are, in fact, the ultimate 'health nuts' – a fact confirmed by medical research findings.

Almonds are so nutritious that the U.S. Department of Agriculture accepts almonds and almond butter as meat alternatives in the National School Lunch Programme.

They are a delicious, nutritious alternative to potato chips or cookies etc. in school lunch boxes, too.

By virtue of a low moisture level, almonds are a concentrated source of nutrition in comparison to higher moisture foods.

An example might help clarify :

– Low nutrient concentration : Milk
– High nutrient concentration: Almonds

Almonds, therefore, are a concentrated source of energy. According to a study, almonds served to delay the onset of fatigue. They help one stay active and alert.

As many as thirteen important vitamins and minerals, essential to the body can be found in almonds. Taken with a balanced diet, almonds promote good health by providing cholesterol-free protein, dietary fibre, calcium, iron, vitamin E. and many other important nutrients.

Almonds are tasty source of complex carbohydrates, the human body's major source of energy.

Protein building blocks that the body cannot manufacture are also contained in almonds.

Almonds are one of the best sources of non-dairy calcium and are recommended for children in their growing years. Almonds are excellent for adolescent girls and women, as well.

They are the best source of iron among popular snacks, and it's a medical requirement that children, adolescents, and women of child-bearing age should consume foods rich in iron.

Almonds are a good source of dietary fibre which promotes a healthy digestive system while also providing protection from many diseases.

Though a significant source of fat, the fat contained in almonds is actually desirable in preventing heart disease. This is because the fat in almonds is monounsaturated or 'good fat' – and this helps reduce blood cholesterol.

Rich in manganese and iron, almonds are excellent for the skin.

Almonds can add nutritive value as well as enhance taste flavour and texture of many dishes when used as an ingredient in cooking, baking, or as toppings and decorative garnishes.

BIBLIOGRAPHY

Amling, H.J., Jacki Small, E. Carden and N. Ronald McDaniel (1982): Nut quality of selected pecan varieties grown in Alabama, Alabama Agril. Expt. Station, Auburn Univ., Alabama Circular 263. pp. 19-24.

Bloch, F. and J.E. Brekke (1960): Processing of pistachio nuts. *Econ. Bot.* 14, 129-144.

Bryant, M.D. (1969): Propagating the pecan trees. New Mex. State Univ. *Pl. Sci. Guide* 400, H-604.

Chandler, W.H. (1957): *Deciduous orchards* (3rd Edition), Lea & Febiger. Philadelphia, pp. 418-457.

Childers, N.F. (1983): *Edible nuts, Minor Tree Crops, Modern Fruit Science.* Horticultural publication 3906 NW 31 Place, Gainesville, Florida, 32606, pp. 283-316.

Commercial Pecan Growing in South Carolina Circular 482. Clemson University Cooperative Extension Service, Clemson, South Carolina.

Diller, J.D., R.B. Clapper and R.A. Jaynes (1964): *Cooperative test plot produce some promising Chinese and hybrid chestnut trees,* U.S. For. Serv. Res. Note, NE-25.

Dimalla, G.G., and J. Van Staden (1978): Pecan nut germination: A review for the nursery industry, *Scient. Hort.* 8 :1-9.

Jacobs, M.B. (1951): *The chemistry and technology of food and food products,* Vol. II, Interscience Publishers, New York.

Jaynes, R.A. (1961): Buried-in-arch technique for rooting chestnut cutting. *Proc. Nor. Nut Grow. Assoc.*, 52, 37-39.

Jaynes, R.A. (1964): Grafting chestnut without stock plants, Proc. Nor. *Nut Grow. Assoc.*, 55, 16-20.

Joley, L.E., and W.E. Whitehouse (1953): Root knot nematodes susceptibility - a factor in the selection of pistachio nut root stocks, *Proc. Amer. Soc. Hort. Sci.*, 61, 99-102.

Lall, H., M.C. Nautiyal and R.M. Sharma (1984): Walnut seed germination-1. Effect of planting depth and seed position in Soil. *Prog. Hort.* 16(2), pp. 6-8.

Menninger, E.A. (1977): *Edible nuts of the world*, Horticultural Books, Inc. P.O. Box 107, Stuart, Florida 33494.

Napier, I., and M. Robbins (1989): *Forest seed and nursery practices in Nepal*, Nepal-U.K. Forestry Research Project, Forestry Res. Div. Deptt. of Forestry and Plant Research, Babar Mahal, P.O. Box 3339, Kathmandu, pp. 90-91.

Nautiyal, M.C. and S.P. Uniyal (1982): Possibilities of walnut cultivation in U.P. hills. *Indian Farmer's Digest*, Vol. XIV, No.10, pp. 27-32.

Noggle, G.R. and G.J. Fritz (1986): *Introductory plant physiology "Plant Growth Substances": Structure and physiological effects*, pp. 448-449, (2nd Edition). Prentice-Hall of India Pvt. Ltd., New Delhi, India.

Peker, K. (1962): Human health and the filbert, a medical history, *Peanut J. and Nut World* 41(5)30-31, 38-40, (6) 38-39.

Sholtto, J.D., Robert A. De J. Hort and G. Shankar Ranganathan (1982): *Forest Farming Prosperity for India*, Natraj Publishers, Dehradun, India.

Singh, R.V., D.C. Khanduri and K. Lal (1973): *Chilgoza pine (Pinus geradiana) regeneration in Himachal Pradesh.* Ind. For., March 1973.

Smith, I.E., B.N. Wolstenholne, and P. Allan (1974): Rooting and establishment of pecan (Carya illinoensis (Wang K. Koch.) stem cuttings, *Agroplantae* 6, 21-28.

Staden, V.J., B.N. Wolstenholne and G.G. Dimalla (1976): Effect of temperatures on pecan seed germination. *Hort. Science* 11(3): 261-262.

Troup, R.S. (1921): *The Silviculture of Indian Trees*, Vol. III, Clarendon Press, Oxford, 1090-1093.

Westwood, M.N. (1978): *Temperate zone pomology*, W.H. Freeman and Company, San Fransisco, pp. 59-65.

Woodroof, J.G. (1979): *Tree Nuts: Production, Processing Products*, AVI Publishing Company Inc, Westport, Connecticut, USA.

Miscellaneous References

University of Arkansas, *Summer care of the pecan orchard*, Cooperative, Extension Service Publication.

Gill and Duffus, Statistics of Hazelnuts, *Edible Nut Statistics.*

Agricultural Products Export Development Authority (APEDA), New Delhi, Product profile of all food-stuff, *Monthly Statistics of Foreign Trade of India.*

Gill and Duffus *Edible Nut Statistics*, Statistics on Almonds.

T.R. Chadha and M.A. Beg, *Walnut : Status and Prospects.*

U.S. Department of Agriculture, Annual Walnut Report of India, Foreign Agricultural Service.

DAVP Publication, *Walnuts from India.*

University of Arkansas establishing pecan trees in Arkansas, Cooperative Extension Service Leaflet 551.

University of Arkansas culture and care of pecan trees in Arkansas, Cooperative Extension Service Leaflet 552.

INDEX